STUDENT'S SOLUTIONS MANUAL

Milton Loyer

Penn State University

to accompany

Essentials of Statistics

Mario F. Triola

Boston San Francisco New York
London Toronto Sydney Tokyo Singapore Madrid
Mexico City Munich Paris Cape Town Hong Kong Montreal

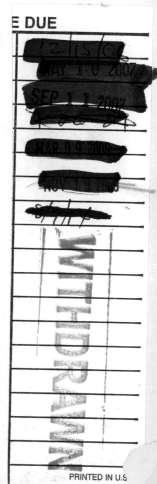

ISBN   0-201-74801-0

1 2 3 4 5 6 7 8 9 10 BB 04 03 02 01

Student Solutions Manual t/a
Triola's *Essentials of Statistics*

Table of Contents

Preface   **Pg. 1**

Chapter 1 Introduction to Statistics   **Pg. 2**

1-2 The Nature of Data   Pg. 2
1-3 Uses and Abuses of Statistics   Pg. 3
1-4 Design of Experiments   Pg. 4
Review Exercises   Pg. 6
Cumulative Review Exercises   Pg. 8

Chapter 2 Describing, Exploring, and Comparing Data   **Pg. 10**

2-2 Summarizing Data with Frequency Tables   Pg. 10
2-3 Pictures of Data   Pg. 14
2-4 Measures of Central Tendency   Pg. 19
2-5 Measures of Variation   Pg. 22
2-6 Measures of Position   Pg. 26
2-7 Exploratory Data Analysis   Pg. 28
Review Exercises   Pg. 32
Cumulative Review Exercises   Pg. 35

Chapter 3 Probability   **Pg. 36**

3-2 Fundamentals   Pg. 36
3-3 Addition Rule   Pg. 38
3-4 Multiplication Rule: Basics   Pg. 40
3-5 Multiplication Rule: Complements and Conditional Probability   Pg. 43
3-6 Counting   Pg. 45
Review Exercises   Pg. 48
Cumulative Review Exercises   Pg. 50

Chapter 4 Probability Distributions   **Pg. 52**

4-2 Random Variables   Pg. 52
4-3 Binomial Experiments   Pg. 54
4-4 Mean, Variance and Standard Deviation for the Binomial Distribution   Pg. 58
Review Exercises   Pg. 60
Cumulative Review Exercises   Pg. 63

Chapter 5 Normal Probability Distributions   **Pg. 63**

5-2 The Standard Normal Distributions   Pg. 63
5-3 Nonstandard Normal Distributions: Finding Probabilities   Pg. 70
5-4 Nonstandard Normal Distributions: Finding Scores   Pg. 74

5-5 The Central Limit Theorem   Pg. 78
5-6 Normal Distribution as Approximation to Binomial Distribution   Pg. 83
Review Exercises   Pg. 86
Cumulative Review Exercises   Pg. 95

Chapter 6 Estimates and Sample Sizes   **Pg. 96**

6-2 Estimating a Population Mean: Large Samples   Pg. 96
6-3 Estimating a Population Mean: Small Samples   Pg. 99
6-4 Determining Sample Size Required to Estimate μ Pg. 101
6-5 Estimating a Population Proportion   Pg. 102
6-6 Estimating a Population Variance   Pg. 106
Review Exercises   Pg. 108
Cumulative Review Exercises   Pg. 110

Chapter 7 Hypothesis Testing   **Pg. 112**

7-2 Fundamentals of Hypothesis Testing   Pg. 112
7-3 Testing a Claim about a Mean: Large Samples   Pg. 114
7-4 Testomg a Claim about a Mean: Small Samples   Pg. 119
7.5 Testing a Claim about a Proportion   Pg. 123
7.6 Testing a Claim about a Standard Deviation or Variance   Pg. 128
Review Exercises   Pg. 130
Cumulative Review Exercises   Pg. 134

Chapter 8 Inferences from Two Samples   **Pg. 136**

8-2 Inferences about Two Means: Independent and Large Samples   Pg. 136
8-3 Inferences about Two Means: Matched Pairs   Pg. 142
8-4 Inferences about Two Proportions   Pg. 146
Review Exercises   Pg. 151
Cumulative Review Exercises   Pg. 154

Chapter 9 Correlation and Regression   **Pg. 156**

9-2 Correlation   Pg. 156
9-3 Regression   Pg. 164
9-4 Variation and Prediction Intervals   Pg. 169
9-5 Rank Correlation   Pg. 172
Review Exercises   Pg. 178
Cumulative Review Exercises   Pg. 183

Chapter 10 Chi-Square and Analysis of Variance   **Pg. 186**

10-2 Multinomial Experiments: Goodness-of-Fit   Pg. 186
10-3 Contingency Tables: Independence and Homogeneity   Pg. 190
10-4 One-Way ANOVA   Pg. 194
Review Exercises   Pg. 200
Cumulative Review Exercises   Pg. 202

PREFACE

This manual contains the solutions to the odd-numbered exercises for each section of the textbook Essentials of Statistics, by Mario Triola, and the solutions for all of the end-of-chapter review and cumulative review exercises of that text. To aid in the comprehension of calculations, worked problems typically include intermediate steps of algebraic and/or computer/calculator notation. When appropriate, additional hints and comments are included and prefaced by NOTE.

Many statistical problems are best solved using particular formats. Recognizing and following these patterns promote understanding and develop the capacity to apply the concepts to other problems. This manual identifies and employs such formats whenever practicable.

For best results, read the text carefully before attempting the exercises, and attempt the exercises before consulting the solutions. This manual has been prepared to provide a check and extra insights for exercises that have already been completed and to provide guidance for solving exercises that have already been attempted but have not been successfully completed.

I would like to thank Mario Triola for writing an excellent elementary statistics book and for inviting me to prepare this solutions manual.

Chapter 1

Introduction to Statistics

1-2 The Nature of Data

1. Statistic, since 20.7 refers to the selected sample.

3. Parameter, since 30 refers to the entire population.

5. Discrete, since the number of absent students must be an integer.

7. Discrete, since the number owning answering machines must be an integer.

9. Ratio, since differences are meaningful and zero height has a natural meaning.

11. Interval, since differences are meaningful but ratios are not. Refer to exercise #19.

13. Interval, since differences are meaningful but ratios are not. Years are not data at the ratio level of measurement because the year zero has been arbitrarily assigned so that the year 0 does not indicate the absence of time. The year 1900, for example, does not represent twice as much time as 950 -- and the ratio would be different using the Chinese or Jewish numerical representations for those years. Since the time difference between 1900 and 1920 is the same as the time difference between 1920 and 1940, however, years are data at the interval level of measurement.

15. Ordinal, since the ratings give relative position in a hierarchy.

17. Ratio, since differences are meaningful and zero ounces has a natural meaning.

19. Temperature ratios are not meaningful because a temperature of 0° does not represent the absence of temperature in the same sense that $0 represents the absence of money. The zero temperature in the example (whether Fahrenheit or Centigrade) was determined by a criterion other than "the absence of temperature."

1-3 Uses and Abuses of Statistics

1. Because the 186,000 respondents were self-selected and not randomly chosen, they are not necessarily representative of the general population and provide no usable information about the general population. In addition, the respondents were self-selected from a particular portion of the general population -- persons watching "Nightline" and able to spend the time and money to respond.

3. a. $500 + (.05)(500) = 500 + 25 = 525$
   b. $525 - (.05)(500) = 525 - 26.25 = 498.75$. No, because the 5% decrease is based on a larger amount than was the previous year's 5% increase.

5. That healthier babies are born to mothers who eat lobsters doesn't mean that eating lobster caused the babies to be healthier. Mothers who eat lobster are probably more affluent than the general population and would tend to eat better, be more knowledgeable about proper pre-natal care, have better health care, etc.

7. Motorcyclists that died in crashes in which helmets may have saved their lives could not be present to testify.

9. There are several possible answers. (1) Since tallness is perceived to be a favorable attribute, people tend to overstate their heights; at the very least, people would tend to round to the next highest inch and not to the nearest inch. (2) Many people do not really accurately know their height. (3) Because Americans tend to express height in feet and inches, errors might occur either in converting heights to all inches or in misstatements like 52" for 5'2". (4) Because many cultures express height in centimeters, some people might not know or be able to readily calculate their heights in inches.

11. No. Since the second 5% price cut would be based on a lower price, two consecutive 5% price cuts yield a smaller price reduction than a single 10% price cut. Mathematically, the two consecutive 5% cuts yield a reduction of $.05x + .05(x-.05x) = .0975x$, or a 9.75% price cut.

13. Assuming that each of the 20 individual subjects is ultimately counted as a success or not (i.e., that there are no "dropouts" or "partial successes"), the success rates in fraction form must be one of 0/20, 1/20, 2/20,..., 19/20, 20/20. In percentages, these rates are multiples of 5 (0%, 5%, 10%,..., 95%, 100%), and values such as 53% and 58% are not mathematical possibilities.

1-4 Design of Experiments

1. Observational study, since specific characteristics are measured on unmodified subjects.

3. Experiment, since the effect of an applied treatment is measured.

5. Random, since each 212 area code telephone number has an equal chance of being selected. But this is really a complex situation, as indicated by the following NOTES.
   NOTE 1: This ignores the fact that some residences may have more than one phone number. A residence with two different phone numbers (e.g., one for the parents and one for the teenagers) has twice the chance of being selected as does a residence with a single phone number.
   NOTE 2: The scenario stated the organization sought to poll "residents" with the 212 area code. If the organization polls all residents at each selected number, this is cluster sampling. If the organization polls one resident at each selected number, the sample is not a random sample of "residents" because a resident living alone and having his own phone number has a higher chance of being selected than a resident living with others (e.g., in a family) and sharing a common phone number.
   NOTE 3: The poll will not include residents in the 212 area code who do not have such phone numbers. This is not a problem if the intended population is phone customers (e.g., for a poll of satisfaction with phone service), but it is if the intended population is general residents (e.g., for a poll of satisfaction with garbage service).

7. Convenience, since the sample was simply those who happen to pass by.

9. Stratified, since the population of interest (assumed to be all car owners) was divided into 5 subpopulations from which the actual sampling was done.

11. Cluster, since the population of interest (assumed to be all students at The College of Newport) was divided into classes which were randomly selected in order to interview all the students in each selected class.
    NOTE: Ideally the division into classes should place each student into one and only one class (e.g., if every student must take exactly one PE class each semester, select the PE classes at random). In practice such divisions are often made in ways that place some students in none of the classes (e.g., by selecting from all 2 pm M-W-F classes) or in more than one of the classes (e.g., by selecting from all the classes offered in the college). With careful handling, imperfect divisions do not significantly affect the results.

13. Stratified, since the population of interest (assumed to be all workers) was divided into 3 subpopulations from which the actual sampling was done.

15. Systematic, since every fifth element in the population (assumed to be all drivers passing the checkpoint during its operation) was sampled.

17. There are several possible answers. (1) Write each full-time student's name on a slip of paper, place the slips in a box, mix them thoroughly, and select 200 of them. (2) Assign each full-time a number (e.g., alphabetically), and use a table of random digits (or a calculator or a computer) to generate 200 random numbers with the appropriate numbers of digits.

19. Obtain from each college bookstore a list of the textbooks currently beings used, and compile a single master list (i.e., without duplications). Number the textbooks on the master list, and use a table of random digits (or a calculator or a computer) to generate 200 numbers with the appropriate numbers of digits.

21. Confounding occurs when the researcher is not able to determine which factor (often one planned and one unplanned) produced an observed effect. If a restaurant tries adding an evening buffet for one week and it is the same week a nearby theater happens to show a real blockbuster that attracts unusual crowds to the neighborhood, the restaurant can not know whether its increased business is due to the new buffet or the extra traffic created by the theater.

23. a. Possibly; no. Stratified random sampling can employ either the same sample size for each stratum or different sample sizes for the various strata. It results in a random sample only when the sample size for each stratum is proportional to the size of the stratum -- i.e., the same <u>proportion</u> (and not the same <u>number</u>) of each stratum is selected for the sample. If the strata are all the same size, then use the same sample size for each; if one stratum is half the size of the others, then its sample size should be half the other sample sizes. If one stratum is half the size of the others and the same sample size is used for each of the strata, then an element in the smaller stratum has a larger chance of being selected than an element in a larger strata -- and that violates the definition of a random sample that requires that each element has the same chance of being selected. Stratified sampling can never result in a simple random sample. It guarantees that the total sample will always include elements from each of the strata, and that total samples without any elements from one of the strata can not occur -- and that violates the definition of simple random sampling that requires that each total sample has the same chance of being selected.

    b. Possibly; no. When each element in the population is in one and only cluster, cluster sampling always results in a random sample. The chance that any element is selected is the chance that its cluster is selected; since each cluster has the same chance of being selected, each element has the same chance of being selected -- and that satisfies the definition of a random sample. Cluster sampling can never result in a simple random sample. It guarantees that total samples with elements from each of the clusters can not occur -- and that violates the definition of simple random sampling that requires that each total sample has the same chance of being selected.

Review Exercises

1. a. Discrete, since the number of shares held must be an integer.
   NOTE: Even if partial shares are allowed (e.g., 5½ shares), the number of shares must be some fractional value and not any value on a continuum -- e.g., a person could not own π shares.
   b. Ratio, since differences between values are consistent and there is a natural zero.
   c. Stratified, since the set of interest (all stockholders) was divided into subpopulations (by states) from which the actual sampling was done.
   d. Statistic, since the value is determined from a sample and not the entire population.
   e. There is no unique correct answer, but the following are reasonable possibilities.
   (1) The proportion of stockholders holding above that certain number of shares (which would vary from company to company) that would make them "influential." (2) The proportion of stockholders holding below that certain number of shares (which would vary from company to company) that would make them "insignificant." (3) The numbers of shares (and hence the degree of influence) held by the largest stockholders.
   f. There are several possible valid answers. (1) The results would be from a self-selected group (i.e., those who chose to respond) and not necessarily a representative group. (2) If the questionnaire did not include information on the numbers of shares owned, the views of small stockholders (who are probably less knowledgeable about business and stocks) could not be distinguished from those of large stockholders (whose views should carry more weight).

2. a. Systematic, since the selections are made at regular intervals.
   b. Convenience, since those selected were the ones who happened to attend.
   c. Cluster, since the stockholders were organized into groups (by stockbroker) and all the stockholders in the selected groups were chosen.
   d. Random, since each stockholder has the same chance of being selected.
   e. Stratified, since the stockholders were divided into subpopulations from which the actual sampling was done.

3. Let N be the total number of full-time students and n be the desired sample size.
   a. Random. Obtain a list of all N full-time students, number the students from 1 to N, select n random numbers from 1 to N, and poll each student whose number on the list is one of the random numbers selected.
   b. Systematic. Obtain a list of all N full-time students, number the students from 1 to N, let m be the largest integer less than the fraction N/n, select a random number between 1 and m, begin with the student whose number is the random number selected, and poll that student and every mth student thereafter.
   c. Convenience. Select a location (e.g., the intersection of major campus walkways) by which most of the students usually pass, and poll the first n full-time students that pass.
   d. Stratified. Obtain a list of all N full-time students and the gender of each, divide the list by gender, and randomly select and poll n/2 students from each gender.
   e. Cluster. Obtain a list of all the classes meeting at a popular time (e.g., 10 am Monday), estimate how many of the classes would be necessary to include n students, select that

many of the classes at random, and poll all of the students in each selected class.

4. a. Blinding occurs when those involved in an experiment (either as subjects or evaluators) do not know whether they are dealing with a treatment or a placebo. It might be used in this experiment by (a) not telling the subjects whether they are receiving Sleepeze or the placebo and/or (b) not telling any post-experiment interviewers or evaluators which subjects received Sleepeze and which ones received the placebo. Double-blinding occurs when neither the subjects nor the evaluators know whether they are dealing with a treatment or a placebo.

   b. The data reported will probably involve subjective assessments (e.g., "On a scale of 1 to 10, how well did it work?") that may be subconsciously influenced by whether the subject was known to have received Sleepeze or the placebo.

   c. In a completely randomized block design, subjects are assigned to the groups (in this case to receive Sleepeze or the placebo) at random.

   d. In a rigorously controlled block design, subjects are assigned to the groups (in this case to receive Sleepeze or the placebo) in such a way that the groups are similar with respect to extraneous variables that might affect the outcome. In this experiment it may be important to make certain each group has approximately the same age distribution, degree of insomnia, number of males, number users of alcohol and/or tobacco, etc.

   e. Replication involves repeating the experiment on a sample of subjects large enough to ensure that atypical responses of a few subjects will not give a distorted view of the true situation.

5. The sample is essentially a convenience sample that might not be representative of the student body. In particular, students likely to drop out may exhibit certain common characteristics (e.g., sleeping in and/or cutting classes) that would make them under-represented in the sample because they would be less likely to pass by the polling location.

6. a. Ratio, since differences are meaningful and zero milligrams of tar has a natural meaning.

   b. Ordinal, since the ratings give relative position in a hierarchy.

   c. Nominal, since the classifications only identify categories and not relative positions on a scale.

   d. Ordinal, since the scores give relative position in a hierarchy but differences are not meaningful -- i.e., the difference in intelligence between IQ's of 40 and 50 is not the same as the difference in intelligence between IQ's of 100 and 110.

   e. Ratio, since differences are meaningful and zero points scored has a natural meaning.

Cumulative Review Exercises

NOTE: Throughout the text intermediate mathematical steps will be shown as an aid to those who may be having difficulty with the calculations. In practice, most of the work can be done continuously on calculators and the intermediate values are unnecessary. Even when the calculations cannot be done continuously, DO NOT WRITE AN INTERMEDIATE VALUE ON YOUR PAPER AND THEN RE-ENTER IT IN THE CALCULATOR. That practice can introduce round-off errors and copying errors. Store any intermediate values in the calculator so that you can recall them with infinite accuracy and without copying errors.

1. $\dfrac{1.23 + 4.56 + 7.89}{3} = (1.23 + 4.56 + 7.89)/3 = 4.56$

2. $\sqrt{\dfrac{(5\text{-}7)^2 + (12\text{-}7)^2 + (4\text{-}7)^2}{3\text{-}1}} = \sqrt{\dfrac{(\text{-}2)^2 + (5)^2 + (\text{-}3)^2}{2}} = \sqrt{\dfrac{4 + 25 + 9}{2}} = \sqrt{\dfrac{38}{2}}$

$$= \sqrt{19} = 4.359$$

3. $\dfrac{1.96^2 \cdot (0.4)(0.6)}{0.025^2} = \dfrac{3.8416 \cdot (.24)}{.000625} = 1475.174$

4. $\dfrac{98.20 - 98.60}{0.62/\sqrt{106}} = \dfrac{\text{-}.40}{.0602} = \text{-}6.642$

5. $\dfrac{25!}{16!9!} = \dfrac{25 \cdot 24 \cdot 23 \cdot 22 \cdot 21 \cdot 20 \cdot 19 \cdot 18 \cdot 17 \cdot 16!}{16!9!} = \dfrac{25 \cdot 24 \cdot 23 \cdot 22 \cdot 21 \cdot 20 \cdot 19 \cdot 18 \cdot 17}{9 \cdot 8 \cdot 7 \cdot 6 \cdot 5 \cdot 4 \cdot 3 \cdot 2 \cdot 1} = 2{,}042{,}975$

NOTE: This exercise should be worked using the factorial key on the calculator. The above intermediate steps illustrate the mathematical logic involved but do not represent an efficient approach to the problem.

6. $\sqrt{\dfrac{10(513.27) - 71.5^2}{10(9)}} = \sqrt{\dfrac{5132.7 - 5112.25}{90}} = \sqrt{\dfrac{20.45}{90}} = \sqrt{.2272} = .477$

7. $\dfrac{8(151{,}879) - (516.5)(2176)}{\sqrt{8(34{,}525.75) - 516.5^2}\,\sqrt{8(728{,}520) - 2176^2}} = \dfrac{1215032 - 1123904}{\sqrt{9433.75}\,\sqrt{1093184}}$

$$= \dfrac{91128}{1015522}$$

$$= .897$$

8. $\dfrac{(183 - 137.09)^2}{137.09} + \dfrac{(30 - 41.68)^2}{41.68} = \dfrac{(45.91)^2}{137.09} + \dfrac{(-11.68)^2}{41.68}$

$= \dfrac{2107.7281}{137.09} + \dfrac{136.4224}{41.68}$

$= 15.375 + 3.273$

$= 18.647$

9. $0.95^{150} = 5.46\text{E-}04$

$= 4.56 \cdot 10^{-4}$

$= .000456$; moving the decimal point left 4 places

NOTE: Calculators and computers vary in their representation of such numbers. This manual assumes they will be given in scientific notation as a two-decimal number between 1.00 and 9.99 inclusive followed by an indication (usually E for *exponent* of the multiplying power of ten) of how to adjust the decimal point to obtain a number in the usual notation (rounded to three significant digits).

10. $25^8 = 1.53\text{E}+11$

$= 1.53 \cdot 10^{11}$

$= 153{,}000{,}000{,}000$; moving the decimal point right 11 places

11. $52^6 = 1.98\text{E}+10$

$= 1.98 \cdot 10^{10}$

$= 19{,}800{,}000{,}000$; moving the decimal point right 10 places

12. $.25^5 = 9.77\text{E-}04$

$= 9.77 \cdot 10^{-4}$

$= .000977$; moving the decimal point left 4 places

Chapter 2

Describing, Exploring, and Comparing Data

2-2 Summarizing Data with Frequency Tables

1. Subtracting two consecutive lower class limits indicates that the class width is 60 - 55 = 5. Since there is a gap of 1.0 between the upper class limit of one class and the lower class limit of the next, class boundaries are determined by increasing or decreasing the appropriate class limits by (1.0)/2 = 0.5. The class boundaries and class midpoints are given in the table below.

height	class boundaries	class midpoint	frequency
55 - 59	54.5 - 59.5	57	1
60 - 64	59.5 - 64.5	62	3
65 - 69	64.5 - 69.5	67	49
70 - 74	69.5 - 74.5	72	46
75 - 79	74.5 - 79.5	77	1
			100

NOTE: Although they often contain extra decimal points and may involve consideration of how the data were obtained, class boundaries are the key to tabular and pictorial data summaries. Once the class boundaries are obtained, everything else falls into place. Here the first class width is readily seen to be 59.5 - 54.5 = 5.0 and the first midpoint is (54.5 + 59.5)/2 = 57. In this manual, class boundaries will typically be calculated first and then used to determine other values. In addition, the sum of the frequencies is an informative number used in many subsequent calculations and will be shown as an integral part of each table.

3. Since the gap between classes as presented is .01, the appropriate class limits are increased or decreased by (.01)/2 = .005 to obtain the class boundaries and the following table.

GPA	class boundaries	class midpoint	frequency
0.00 - 0.49	-0.005 - 0.495	0.245	72
0.50 - 0.99	0.495 - 0.995	0.745	23
1.00 - 1.49	0.995 - 1.495	1.245	47
1.50 - 1.99	1.495 - 1.995	1.745	135
2.00 - 2.49	1.995 - 2.495	2.245	288
2.50 - 2.99	2.495 - 2.995	2.745	276
3.00 - 3.49	2.995 - 3.495	3.245	202
3.50 - 3.99	3.495 - 3.995	3.745	97
			1140

The class width is 0.495 - (-0.005) = .50; the first midpoint is (-0.005 + 0.495)/2 = 0.245.

5. The relative frequency for each class is found by dividing its frequency by 100, the sum of the frequencies. NOTE: As before, the sum is included as an integral part of the table. For relative frequencies, this should always be 1.000 (i.e., 100%) and serves as a check for the calculations.

height	relative frequency
55 - 59	.01
60 - 64	.03
65 - 69	.49
70 - 74	.46
74 - 79	.01
	1.00

7. The relative frequency for each class is found by dividing its frequency by 1140, the sum of the frequencies. NOTE: In #5, the relative frequencies were expressed as decimals; here they are expressed as percents. The choice is arbitrary.

GPA	relative frequency
0.00 - 0.49	6.32%
0.50 - 0.99	2.02%
1.00 - 1.49	4.12%
1.50 - 1.99	11.84%
2.00 - 2.49	25.26%
2.50 - 2.99	24.21%
3.00 - 3.49	17.72%
3.50 - 3.99	8.51%
	100.00%

9. The cumulative frequencies are determined by repeated addition of successive frequencies to obtain the combined number in each class and all previous classes. NOTE: Consistent with the emphasis that has been placed on class boundaries, we choose to use upper class boundaries in the "less than" column. Conceptually, heights occur on a continuum and the integer values reported are assumed to be the nearest whole number representation of the precise measure of height. An exact height of 59.7, for example, would be reported as 60 and fall in the second class. The values in the first class, therefore, are better described as being "less than 59.5" (using the upper class boundary) than as being "less than 60." This distinction becomes crucial in the construction of pictorial representations in the next section. In addition, the fact that the final cumulative frequency must equal the total number (i.e, the sum of the frequency column) serves as a check for calculations. The sum of cumulative frequencies, however, has absolutely no meaning and is not included.

height	cumulative frequency
less than 59.5	1
less than 64.5	4
less than 69.5	53
less than 74.5	99
less than 79.5	100

11. The cumulative frequencies are determined by repeated addition of successive frequencies to obtain the combined number in each class and all previous classes. NOTE: Consistent with the emphasis that has been placed on class boundaries, we choose to use upper class boundaries in the "less than" column.

GPA	cumulative frequency
less than 0.495	72
less than 0.995	95
less than 1.495	142
less than 1.995	277
less than 2.495	565
less than 2.995	841
less than 3.495	1043
less than 3.995	1140

13. Assuming that "start the first class at 0.7900 lb" refers to the first lower class limit produces the frequency table below and violates none of the guidelines for constructing frequency tables.

weight (lbs)	frequency
.7900 - .7949	1
.7950 - .7999	0
.8000 - .8049	1
.8050 - .8099	3
.8100 - .8149	4
.8150 - .8199	17
.8200 - .8249	6
.8250 - .8299	4
	36

NOTE: The class boundaries above are .78995, .79495, .79995, etc. Using 0.7900 as the first lower class boundary produces boundaries of .7900, .7950, .8000, etc. This is not acceptable, as these are possible data values. This introduces subjectivity about where to place a value that falls on the boundary and violates the guideline that each of the values must belong to only one class.

14. Assuming that "start the first class at 0.7750 lb" refers to the first lower class limit produces the frequency table below and violates the guideline that frequency tables should have between 5 and 20 classes.

weight (lbs)	frequency
.7750 - .7799	4
.7800 - .7849	13
.7850 - .7899	15
.7900 - .7949	4
	36

NOTE: That this frequency table has only 4 categories, which is usually not sufficient to give a picture of the nature of the distribution, is allowable in this context -- since the class limits employed work well with the other cola data and permit meaningful comparisons across the data sets.

15. Assuming that "start the first class at 0.8100 lb" refers to the first lower class limit produces the frequency table below.

weight (lbs)	frequency
.8100 - .8149	1
.8150 - .8199	6
.8200 - .8249	16
.8250 - .8299	8
.8300 - .8349	3
.8350 - .8399	1
.8400 - .8449	1
	36

While similar to the frequency table in exercise #13, this table differs in two ways. (1) In exercise #13 [Coke], there were 5 classes below the modal class and 2 above; in exercise #15 [Pepsi], there are 2 classes below the modal class and 4 above. (2) The weights in exercise #13 appear to be less than those in exercise #15.

16. Assuming that "start the first class at 0.7700 lb" refers to the first lower class limit produces the frequency table below.

weight (lbs)	frequency
.7700 - .7749	1
.7750 - .7799	6
.7800 - .7849	14
.7850 - .7899	13
.7900 - .7949	2
	36

While similar to the frequency table in exercise #15, this table differs in two ways. (1) In exercise #15 [regular Pepsi], there were 2 classes below the modal class and 4 above; in exercise #16 [diet Pepsi], there are 2 classes below the modal class and 4 above. In both cases, however, there are more values above the modal class than below it. (2) The weights in exercise #15 appear to be greater than those in exercise #16.

17. For 11 classes to cover data ranging from
a beginning lower class limit of 0 to a maximum
value of 514, the class width must be at least
(514 - 0)/11 = 46.7. A convenient class
width would be 50, which produces the frequency
table given at the right.

weight (lbs)	frequency
00 - 49	6
50 - 99	10
100 - 149	10
150 - 199	7
200 - 249	8
250 - 299	2
300 - 349	4
350 - 399	3
400 - 449	3
450 - 499	0
500 - 549	1
	36

19. Assuming that "start the first class at 200 lb" refers to the first lower class limit produces
the frequency table below.

weight (lbs)	frequency
200 - 219	12
220 - 239	9
240 - 259	18
260 - 279	84
280 - 299	52
	175

Yes. Since the lowest recorded weight before collapse is over 200 and most of the weights
are over 260, the cans should withstand a pressure that varies between 158 and 165.

20. Assuming that "start the first class at 200 lb" refers to the first lower class limit produces
the frequency table given at the right in exercise #21.

Yes. Most of the thicker cans support a weight of 280 before collapse, and they appear to
be stronger. Since the thinner cans already meet the criterion given in exercise #19,
however, the added strength of the thicker cans may not be worth the added cost.

21. Assuming that "start the first class at 200 lb" refers to the first lower class limit produces
the frequency table below.

weight (lbs)	frequency
200 - 219	6
220 - 239	5
240 - 259	12
260 - 279	36
280 - 299	87
300 - 319	28
320 - 339	0
340 - 359	0
360 - 379	0
380 - 399	0
400 - 419	0
420 - 439	0
440 - 459	0
460 - 479	0
480 - 499	0
500 - 519	1
	175

[table for exercise #20]

weight (lbs)	frequency
200 - 219	6
220 - 239	5
240 - 259	12
260 - 279	36
280 - 299	87
300 - 319	28
	174

In general, an outlier can add several rows to a frequency table. Even though most of the
added rows have frequency zero, the table tends to suggest that these are possible valid
values – thus distorting the reader's mental image of the distribution.

2-3 Pictures of Data

1. 42, the height of the bar centered at 0.0

3. 2, the one Monday represented by the bar centered at 1.0 and the one Monday represented by the bar centered at 1.4

5. See the figure below. The bars extend from class boundary to class boundary. Each axis is labeled numerically <u>and</u> with the name of the quantity represented. Barring an interval longer than any previously recorded, a minimum stay of 109.5 minutes assures seeing an eruption. A minimum stay of 99.5 minutes includes 199 of the 200 (i.e. 99.5%) recorded intervals and would be inadequate only if the longest interval occurred *and* the tour arrived within the first ten minutes of that interval -- or about $(1/200) \times (10/110) = 0.00045 = .045\%$ of the time, about once in every 2200 tours.

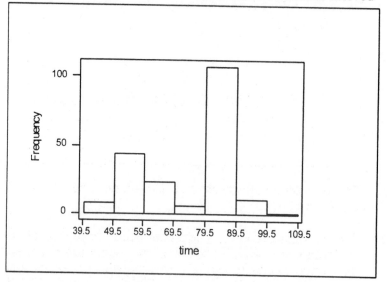

7. See the figure below. The bars extend from class boundary to class boundary. Each axis is labeled numerically <u>and</u> with the name of the quantity represented. Although the posted limit is 30 mph, it appears that the police ticket only those traveling at least 42 mph.

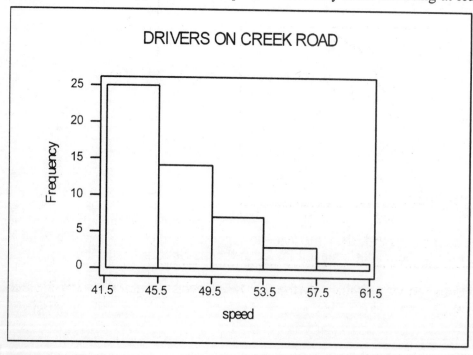

9. See the figure below. The frequencies are plotted above the class midpoints, and "extra" midpoints are added so that both polygons begin and end with a frequency of zero. The horizontal axis is labeled for convenience of presentation, but the frequencies are plotted beginning at .77245 and every .005 thereafter − .77245, .77745,…,.82745,.83245.

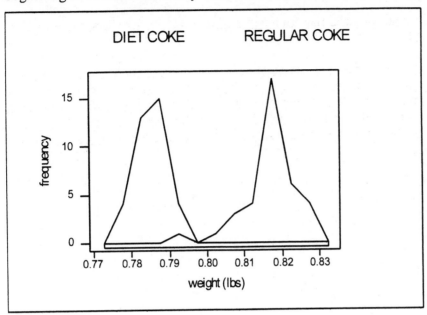

The two distributions have similar shapes, but the diet Coke weights are less than those of regular Coke.

11. See the figure below. The frequencies are plotted above the class midpoints, and "extra" midpoints are added so that both polygons begin and end with a frequency of zero. The horizontal axis is labeled for convenience of presentation, but the frequencies are plotted beginning at .76745 and every .005 thereafter − .76745, .77245,…,.79245,.79745.

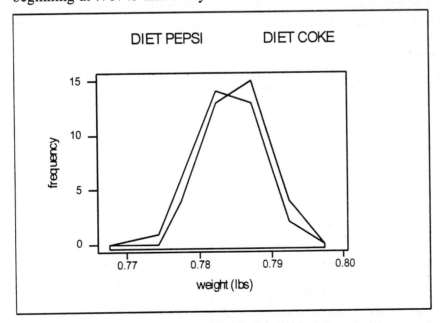

The two distributions are very similar, but the diet Pepsi weights are slightly smaller than those of diet Coke.

13. The original numbers are listed by the row in which they appear in the stem-and-leaf plot.

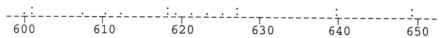

stem	leaves	original numbers
60	0117	600, 601, 601, 607
61	02889	610, 612, 618, 618, 619
62	13577	621, 623, 625, 627, 627
63		
64	0099	640, 640, 649, 649

15. The dotplot is constructed using the original scores as follows. Each space represents 1 unit.

```
    :                   :   :                   :       :
--+------------.---------+---.-:-:-:----------+----------+--
  600        610       620       630        640       650
```

17. The expanded stem-and-leaf plot below on the left is one possibility. NOTE: The text claims that stem-and-leaf plots enable us to "see the distribution of data and yet keep all the information in the original list." Following the suggestion to round the nearest inch not only loses information but also uses subjectivity to round values exactly half way between. Since always rounding such values "up" creates a slight bias, many texts suggest rounding toward the even digit -- so that 33.5 becomes 34, but 36.5 becomes 36. The technique below of using superscripts to indicate the occasional decimals is both mathematically clear and visually uncluttered.

stem	leaves
3	6 7
4	0 0 1 3 3^5
4	6 6 7 8 8 9
5	0 2 2^5 3 3 4
5	7^3 7^5 8 9 9 9
6	0 0^5 1 1 1^5 2 3 3 3 3^5 4 4 4
6	5 5 6^5 7 7^5 8^5
7	0 0^5 2 2 2 2 3 3^5
7	5 6^5

19. See the figure below, with bars arranged in order of magnitude. Networking appears to be the most effective job-seeking approach.

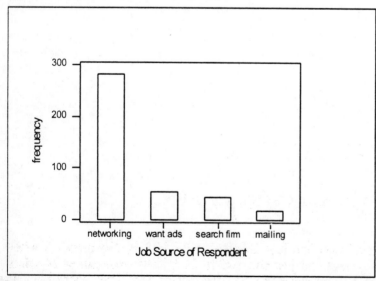

21. See the figure below. The sum of the frequencies is 50; the relative frequencies are 23/50 = 46%, 9/50 = 18%, 12/50 = 24%, and 6/50 = 12%. The corresponding central angles are (.46)360° = 165.6°, (.18)360° = 64.8°, (.24)360° = 86.4°, and (.12)360° = 43.2°. To be complete, the figure needs to be titled with the name of the quantity being measured.

Job Source of Responder

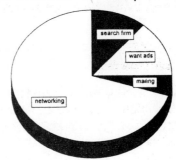

23. The scatter diagram is given below. The figure should have a title, and each axis should be labeled both numerically and with the name of the variable. An "x" marks a single occurrence, while numbers indicate multiple occurrences at a point. Cigarettes high in tar also tend to be high in CO. The points cluster about a straight line from (0,0) to (18,18), indicating that the mg of CO tends to be about equal to the mg of tar.

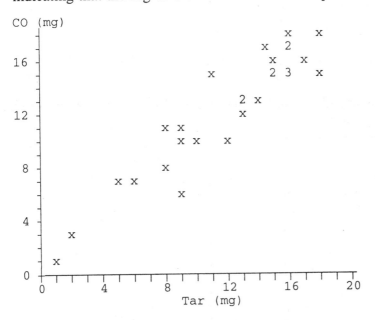

TAR AND CO CONTENT OF SELECTED 100mm FILTERED AMERICAN CIGARETTES

25. According to the figure, 422,000 started and 10,000 returned. 10,000/422,000 = 2.37%

27. The figure indicates the number of men had just dropped to 37,000 on November 9 when the temperature was 16°F (-9°C), and had just dropped to 24,000 on November 14 when the temperature was -6°F (-21°C). The number who died during that time, therefore, was 37,000 - 24,000 = 13,000.

29. NOTE: Exercises #20 and #21 of section 2-2 dealt with frequency table representations of this data and specified a class width of 20. Using a class width with an odd number of units of measure allows the class midpoint to have the same number of decimal places as the original data and often produces more appealing visual representations. Here we use a class width of 25 – with class midpoints of 200, 225, etc. The histogram bars extend from class boundary to class boundary (i.e., from 187.5 to 212.5 for the first class), but for convenience the labels have been placed at the class midpoints.

a. The histogram with the outlier is as shown below.

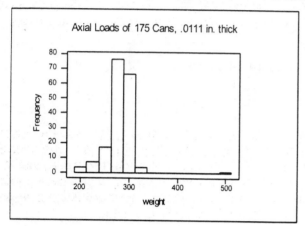

b. The histogram without the outlier is as shown below.

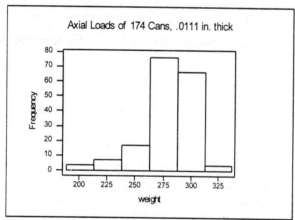

c. The basic shape of the histogram does not change, except that a distant piece has been "broken off." NOTE: The redrawn histogram in part (b) should not be an exact copy of the one in part (a) with the distant bar erased. Since removing the distant bar reduces the effective width of the figure significantly, the rule that the height should be approximately 3/4 of the width requires either making the remaining bars wider (and keeping the figure's height) or making them shorter (and keeping the figure's reduced width). To do otherwise produces a figure too tall for its width -- and one that tends to visually overstate the differences between classes.

2-4 Measures of Central Tendency

NOTE: As it is common in mathematics and statistics to use symbols instead of words to represent quantities that are used often and/or that may appear in equations, this manual employs symbols for the measures of central tendency as follows:

mean = $\bar{x}$ mode = M

median = $\tilde{x}$ midrange = m.r.

Also, this manual follows the author's guideline of presenting means, medians and ranges accurate to one more decimal place than found in the original data. The mode, the only measure which must be one of the original pieces of data, is presented with the same accuracy as the original data.

1. Arranged in order, the 12 scores are: 8 11 13 14 14 14 15 16 17 18 25 27
   a. $\bar{x} = (\Sigma x)/n = (192)/12 = 16.0$ c. M = 14
   b. $\tilde{x} = (14 + 15)/2 = 14.5$ d. m.r. = (8 + 27)/2 = 17.5

NOTE: The median is the middle score <u>when the scores are arranged in order</u>, and the midrange is halfway between the first and last score <u>when the scores are arranged in order</u>. It is therefore usually helpful to begin by placing the scores in order. This will not affect the mean, and it may also aid in identifying the mode. In addition, no measure of central tendency can have a value lower than the smallest score or higher than the largest score -- remembering this helps to protect against gross errors, which most commonly occur when calculating the mean.

3. Arranged in order, the 12 scores are: 35 46 55 65 74 83 88 93 99 107 108 119
   a. $\bar{x} = (\Sigma x)/n = (972)/12 = 81.0$ c. M = [none]
   b. $\tilde{x} = (83 + 88)/2 = 85.5$ d. m.r. = (35 + 119)/2 = 77.0

   Yes; these results are acceptable and consistent with the goal of service in 90 seconds or less.

5. Arranged in order, the scores are as follows.
   JV: 6.5 6.6 6.7 6.8 7.1 7.3 7.4 7.7 7.7 7.7
   Pr: 4.2 5.4 5.8 6.2 6.7 7.7 7.7 8.5 9.3 10.0

   <u>Jefferson Valley</u>   <u>Providence</u>
   n = 10   n = 10
   $\bar{x} = (\Sigma x)/n = (71.5)/10 = 7.15$ $\bar{x} = (\Sigma x)/n = (71.5)/10 = 7.15$
   $\tilde{x} = (7.1 + 7.3)/2 = 7.20$ $\tilde{x} = (6.7 + 7.7)/2 = 7.20$
   M = 7.7   M = 7.7
   m.r. = (6.5 + 7.7)/2 = 7.10   m.r. = (4.2 + 10.0)/2 = 7.10

   Comparing only measures of central tendency, one might suspect the two sets are identical. The Jefferson Valley times, however, are considerably less variable.
   NOTE: This is the reason most banks have gone to the single waiting line. While it doesn't make service faster, it makes service times more equitable by eliminating the "luck of the draw" -- i.e., ending up by pure chance in a fast or slow line and having unusually short or long waits.

7. Arranged in order, the scores are as follows.
   Coke: .8150 .8163 .8181 .8192 .8211 .8247
   Pepsi: .8156 .8170 .8211 .8216 .8258 .8302

   <u>Coke</u>   <u>Pepsi</u>
   n = 6   n = 6
   $\bar{x} = (\Sigma x)/n = (4.9144)/6 = .81907$ $\bar{x} = (\Sigma x)/n = (4.9313)/6 = .82188$
   $\tilde{x} = (.8181 + .8192)/2 = .81865$ $\tilde{x} = (.8211 + .8216)/2 = .82135$
   M = [none]   M = [none]
   m.r. = (.8150 + .8247)/2 = .81985   m.r. = (.8156 + .8302)/2 = .82290

Pepsi appears to weigh slightly more than Coke. Since the cans are sold by volume and not by weight, this is not a question of ethics -- merely a reflection of the fact that the two drink cans of the same volume have different ingredients.

9. The x values below are the class midpoints from the given frequency table.

x	f	x·f
44.5	8	356.0
54.5	44	2398.0
64.5	23	1483.5
74.5	6	447.0
84.5	107	9041.5
94.5	11	1039.5
104.5	1	104.5
	200	14870.0

$$\overline{x} = (\Sigma x \cdot f)/n$$
$$= (14870.0)/200$$
$$= 74.35$$

NOTE: The mean time was calculated to be 74.35 minutes. According to the rule given in the text, this value should be rounded to one decimal place. The text describes how many decimal places to present in an answer, but not the actual rounding process. When the figure to be rounded is <u>exactly</u> half-way between two values (i.e., the digit in the position to be discarded is a 5, and there are no further digits because the calculations have "come out even"), there is no universally accepted rounding rule. Some authors say to always round up such a value; others correctly note that always rounding up introduces a consistent bias, and that the value should actually be rounded up half the time and rounded down half the time. And so some authors suggest rounding toward the even value (e.g., .65 becomes .6 and .75 becomes .8), while others simply suggest flipping a coin. In this manual, answers <u>exactly</u> half-way between will be reported without rounding (i.e., stated to one more decimal than usual).

11. The x values below are the class midpoints from the given frequency table.

x	f	x·f
43.5	25	1087.5
47.5	14	665.0
51.5	7	360.5
55.5	3	166.5
59.5	1	59.5
	50	2339.0

$$\overline{x} = (\Sigma x \cdot f)/n$$
$$= (2339.0)/50$$
$$= 46.8$$

The mean speed of 46.8 mi/hr of those ticketed by the police is more than 1.5 times the posted speed limit of 30 mi/hr. NOTE: This indicates nothing about the mean speed of *all* drivers, a figure which may or may not be higher than the posted limit.

13. The following values were obtained for new textbooks, where x_i indicates the ith score from the ordered list.

author's college	UMASS
n = 35	n = 40
$\overline{x} = (\Sigma x)/n = 2016.85/35 = 57.624$	$\overline{x} = (\Sigma x)/n = 2607.70/40 = 65.1175$
$\tilde{x} = x_{18} = 59.35$	$\tilde{x} = (x_{20}+x_{21})/2 = (70.00 + 72.70)/2 = 71.35$

Assuming a random sample was taken from each institution, textbooks appear to cost more at the University of Massachusetts than at the author's college. This is probably more a reflection of the courses offered at each institution and/or the textbook selection practices of faculty than of bookstore pricing policies.

15. The following values were obtained for Boston rainfall, where x_i indicates the ith score from the ordered list.

Thursday
$n = 52$
$\bar{x} = (\Sigma x)/n = 3.57/52 = .069$
$\tilde{x} = (x_{26}+x_{27})/2 = (.00 + .00)/2 = .000$

Sunday
$n = 52$
$\bar{x} = (\Sigma x)/n = 3.52/52 = .068$
$\tilde{x} = (x_{26}+x_{27})/2 = (.00 + .00)/2 = .000$

If "it rains more on weekends" refers to the amount of rain, the data do not support the claim. The amount of rainfall appears to be virtually the same for Thursday and Sunday. If "it rains more on weekends" refers to the frequency of rain (regardless of the amount), then the proportions of days on which there was rain would have to be compared.

17. Let $\bar{x}_h$ stand for the harmonic mean: $\bar{x}_h = n/[\Sigma(1/x)]$
$= 2/[1/40 + 1/60]$
$= 2/[.0417]$
$= 48.0$

19. R.M.S. $= \sqrt{\Sigma x^2/n}$
$= \sqrt{[(110)^2 + (0)^2 + (-60)^2 + (12)^2]/4}$
$= \sqrt{15844/4}$
$= \sqrt{3961}$
$= 62.9$

21. a. Arranged in order, the original 54 scores are:
```
26   29   34   40   46   48   60   62   64   65   76   79   80   86   90
94  105  114  116  120  125  132  140  140  144  148  150  150  154  166
166  180  182  202  202  204  204  212  220  220  236  262  270  316  332
344  348  356  360  365  416  436  446  514
```
$\bar{x} = (\Sigma x)/n = (9876)/54 = 182.9$

b. Trimming the highest and lowest 10% (or 5.4 = 5 scores), the remaining 44 scores are:
```
48   60   62   64   65   76   79   80   86   90   94  105  114  116  120
125  132  140  140  144  148  150  150  154  166  166  180  182  202  202
204  204  212  220  220  236  262  270  316  332  344  348  356  360
```
$\bar{x} = (\Sigma x)/n = (7524)/44 = 171.0$

c. Trimming the highest and lowest 20% (or 10.8 = 11 scores), the remaining 32 scores are:
```
79   80   86   90   94  105  114  116  120  125  132  140  140  144  148
150  150  154  166  166  180  182  202  202  204  204  212  220  220  236
262  270
```
$\bar{x} = (\Sigma x)/n = (5093)/32 = 159.2$

In this case, the mean gets smaller as more scores are trimmed. In general, means can increase, decrease, or stay the same as more scores are trimmed. The mean decreased here because the higher scores were farther from the original mean than were the lower scores.

2-5 Measures of Variation

NOTE: Although not given in the text, the symbol R will be used for the range throughout this manual. Remember that the range is the difference between the highest and the lowest scores, and not necessarily the difference between the last and the first values as they are listed. Since calculating the range involves only the subtraction of 2 original pieces of data, that measure of variation will be reported with the same accuracy as the original data.

1.

x	x $-\bar{x}$	$(x-\bar{x})^2$	x^2
8	-8	64	64
11	-5	25	121
13	-3	9	169
14	-2	4	196
14	-2	4	196
14	-2	4	196
15	-1	1	225
16	0	0	256
17	1	1	289
18	2	4	324
25	9	81	625
27	11	121	729
192	0	318	3390

$\bar{x} = (\Sigma x)/n = 192/12 = 16.0$

$R = 27 - 8 = 19$

by formula 2-4,
$s^2 = \Sigma(x-\bar{x})^2/(n-1)$
   $= 318/11$
   $= 28.9091$
   $= 28.9$

by formula 2-5,
$s^2 = [n(\Sigma x^2) - (\Sigma x)^2]/[n(n-1)]$
   $= [12(3390) - (192)^2]/[12(11)]$
   $= [3816]/[132]$
   $= 28.9091$
   $= 28.9$

$s = \sqrt{28.9091} = 5.4$

NOTE: When finding the square root of the variance to obtain the standard deviation, use all the decimal places of the variance, and not the rounded value reported as the answer. The best way to do this is either to keep the value on the calculator display or to place it in the memory. Do not copy down all the decimal places and then re-enter them to find the square root, as that could introduce round-off and/or copying errors.

When using formula 2-4, constructing a table having the first three columns shown above helps to organize the calculations and makes errors less likely. In addition, verify that $\Sigma(x-\bar{x}) = 0$ before proceeding -- if such is not the case, there is an error and further calculation is fruitless. For completeness, and as a check, both formulas 2-4 and 2-5 were used above. In general, only formula 2-5 will be used throughout the remainder of this manual for the following reasons:
   (1) When the mean does not "come out even," formula 2-4 involves round-off error and/or many messy decimal calculations.
   (2) The quantities Σx and Σx^2 needed for formula 2-5 can be found directly and conveniently on the calculator from the original data without having to construct a table like the one above.

3. preliminary values: n = 121, $\Sigma x = 972$, $\Sigma x^2 = 86424$
    R = 119 - 35 = 84
    $s^2 = [n(\Sigma x^2) - (\Sigma x)^2]/[n(n-1)]$
      $= [12(86424) - (972)^2]/[12(11)]$
      $= (92304)/132$
      $= 699.3$
    s = 26.4

NOTE: The quantity $[n(\Sigma x^2) - (\Sigma x)^2]$ cannot be less than zero. A negative value indicates that there is an error and that further calculation is fruitless. In addition, remember to find the value for s by taking the square root of the precise value of s^2 showing on the calculator display before it is rounded to one more decimal place than the original data.

5. Jefferson Valley

$n = 10, \Sigma x = 71.5, \Sigma x^2 = 513.27$

$R = 7.7 - 6.5 = 1.2$

$s^2 = [n(\Sigma x^2) - (\Sigma x)^2]/[n(n-1)]$

$\quad = [10(513.27) - (71.5)^2]/[10(9)]$

$\quad = 20.45/90$

$\quad = 0.23$

$s = 0.48$

Providence

$n = 10, \Sigma x = 71.5, \Sigma x^2 = 541.09$

$R = 10.0 - 4.2 = 5.8$

$s^2 = [n(\Sigma x^2) - (\Sigma x)^2]/[n(n-1)]$

$\quad = [10(541.09) - (71.5)^2]/[10(9)]$

$\quad = 298.65/90$

$\quad = 3.32$

$s = 1.82$

Exercise #5 of section 2-4 indicated that the mean waiting time was 7.15 minutes at each bank. The Jefferson Valley waiting times, however, are considerably less variable. The range measures the difference between the extremes. The longest and shortest waits at Jefferson Valley differ by a little over 1 minute (R=1.2), while the longest and shortest waits at Providence differ by almost 6 minutes (R=5.8). The standard deviation measures the typical difference from the mean. A Jefferson Valley customer usually receives service within about ½ minute (s=0.48) of 7.15 minutes, while a Providence customer usually receives service within about 2 minutes (s=1.82) of the mean.

7. Coke

$n = 6, \Sigma x = 4.9144, \Sigma x^2 = 4.02528224$

$R = .8247 - .8150 = .0097$

$s^2 = [n(\Sigma x^2) - (\Sigma x)^2]/[n(n-1)]$

$\quad = [6(4.02528224) - (4.9144)^2]/[6(5)]$

$\quad = .00036608/30$

$\quad = .000012202$

$s = .00349$

Pepsi

$n = 6, \Sigma x = 4.9313, \Sigma x^2 = 4.05310181$

$R = .8302 - .8156 = .0146$

$s^2 = [n(\Sigma x^2) - (\Sigma x)^2]/[n(n-1)]$

$\quad = [6(4.05310181) - (4.9313)^2]/[6(5)]$

$\quad = .00089117/30$

$\quad = .000029706$

$s = .00545$

There appears to be slightly more variation among the Pepsi weights than among those for Coke. In other words, Coke seems to be doing slightly better when it comes to producing a uniform product.

NOTE: Following the usual round-off rule of giving answers with one more decimal than the original data produces variances of .00001 and .00003, which have only one significant digit. This is an unusual situation occurring because values less than 1.0 become smaller when they are squared. Since the original data had 4 significant digits, we provide 5 significant digits for the variance.

9.

x	f	f·x	f·x²
44.5	8	356.0	15842.00
54.5	44	2398.0	130691.00
64.5	23	1483.5	95685.75
74.5	6	447.0	33301.50
84.5	107	9041.5	764006.75
94.5	11	1039.5	98232.75
104.5	1	104.5	10920.25
	200	14870.0	1148680.00

$s^2 = [n(\Sigma f \cdot x^2) - (\Sigma f \cdot x)^2]/[n(n-1)]$

$\quad = [200(1148680.00) - (1487.0)^2]/[200(199)]$

$\quad = (8619100.00)/29800$

$\quad = 216.56$

$s = 14.7$

11.

x	f	f·x	f·x²
43.5	25	1087.5	47305.25
47.5	14	665.0	31587.50
51.5	7	360.5	18565.75
55.5	3	166.5	9240.75
59.5	1	59.5	3540.25
	50	2339.0	110239.50

$s^2 = [n(\Sigma f \cdot x^2) - (\Sigma f \cdot x)^2]/[n(n-1)]$
$= [50(110239.50) - (2339.0)^2]/[50(49)]$
$= (41054.00)/2450$
$= 16.76$
$s = 4.1$

13. author's college
    $n = 35, \Sigma x = 2016.85, \Sigma x^2 = 123549.5808$
    $s^2 = [n(\Sigma x^2) - (\Sigma x)^2]/[n(n-1)]$
    $= [35(123549.5808) - (2016.85)^2]/[35(34)]$
    $= 256551.4055/1190$
    $= 215.5894$
    $s = 14.683$

    UMASS
    $n = 40, \Sigma x = 2604.70, \Sigma x^2 = 190378.5631$
    $s^2 = [n(\Sigma x^2) - (\Sigma x)^2]/[n(n-1)]$
    $= [40(190378.5631) - (2604.70)^2]/[40(39)]$
    $= 830680.4340/1560$
    $= 532.4875$
    $s = 23.076$

    There appears to be more variation among prices of new textbooks at UMASS than at the author's college.

15. Thursday
    $n = 52, \Sigma x = 3.57, \Sigma x^2 = 1.6699$
    $s^2 = [n(\Sigma x^2) - (\Sigma x)^2]/[n(n-1)]$
    $= [52(1.6699) - (3.57)^2]/[52(51)]$
    $= 74.0899/2652$
    $= .02793$
    $s = .167$

    Sunday
    $n = 52, \Sigma x = 3.52, \Sigma x^2 = 2.2790$
    $s^2 = [n(\Sigma x^2) - (\Sigma x)^2]/[n(n-1)]$
    $= [52(2.2790) - (3.52)^2]/[52(51)]$
    $= 106.1176/2652$
    $= .04001$
    $s = .200$

    The values are close, although there may be slightly more variation among the amounts for Sundays than among the amounts for Thursdays.

17. Assuming that the graduates ranged in age from 17 to 19, the *Range Rule of Thumb* suggests $s \approx range/4 = (19-17)/4 = 2/4 = 0.5$.

19. Given $\overline{x} = 75$ and $s = 12$, the Range Rule of Thumb suggests
    minimum "usual" value $= \overline{x} - 2s = 75 - 2(12) = 75 - 24 = 51$
    maximum "usual" value $= \overline{x} + 2s = 75 + 2(12) = 75 + 24 = 99$
    Yes, in this context a score of 50 would be considered unusually low.

21. a. The limits 61.1 and 66.1 are 1 standard deviation from the mean. The *Empirical Rule for Data with a Bell-shaped Distribution* states that about 68% of the heights should fall within those limits.
    b. The limits 56.1 and 76.1 are 3 standard deviations from the mean. The *Empirical Rule for Data with a Bell-shaped Distribution* states that about 99.7% of the heights should fall within those limits.

23. The limits 58.6 and 68.6 are 2 standard deviations from the mean. *Chebyshev's Theorem* states that there must be at least $1 - 1/k^2$ of the scores within k standard deviations of the mean. Here k = 2, and so the proportion of the heights within those limits is at least $1 - 1/2^2 = 1 - 1/4 = 3/4 = 75\%$.

25. A standard deviation of s = 0 is possible only when $s^2 = 0$, and $s^2 = \Sigma(x-\overline{x})^2/(n-1) = 0$ only when $\Sigma(x-\overline{x})^2 = 0$. Since each $(x-\overline{x})^2$ is non-negative, $\Sigma(x-\overline{x})^2 = 0$ only when every $(x-\overline{x})^2 = 0$ – i.e., only when every x is equal to $\overline{x}$. In simple terms, no variation occurs only when all the scores are identical.

27. The Everlast brand is the better choice. In general, a smaller standard deviation of lifetimes indicates more consistency from battery to battery – signaling a more dependable production process and a more dependable final product. Assuming a bell-shaped distribution of lifetimes, for example, that empirical rule states that about 68% of the lifetimes will fall within one standard deviation of the mean. Here, those limits would be

for Everlast: 50 $\pm$ 2 or 48 months to 52 months
for Endurance: 50 $\pm$ 6 or 44 months to 56 months

While a person might be lucky and purchase a long-lasting Endurance battery, an Everlast battery is much more likely to last for the advertised 48 months.

29. section 1   section 2

section 1	section 2
$n = 11$, $\Sigma x = 201$, $\Sigma x^2 = 4001$	$n = 11$, $\Sigma x = 119$, $\Sigma x^2 = 1741$
$R = 20 - 1 = 19$	$R = 19 - 2 = 17$
$s^2 = [n(\Sigma x^2) - (\Sigma x)^2]/[n(n-1)]$	$s^2 = [n(\Sigma x^2) - (\Sigma x)^2]/[n(n-1)]$
$\quad = [11(4001) - (201)^2]/[11(10)]$	$\quad = [11(1741) - (119)^2]/[11(10)]$
$\quad = 3610/110$	$\quad = 1990/110$
$\quad = 32.92$	$\quad = 45.36$
$s = 5.7$	$s = 6.7$

The range values give the impression that section 1 had more variability than section 2. The range can be misleading because it is based only on the extreme scores. In this case, the lowest score in section 1 was so distinctly different from the others that to include it in any measure trying to give a summary about the section as a whole would skew the results. For the mean, where the value is only one of 11 used in the calculation, the effect is minimal; for the range, where the value is one of only 2 used in the calculation, the effect is dramatic. The standard deviation values give the impression that section 2 had slightly more variability.

NOTE: In this case, section 2 seems considerably more variable (or diverse), and even the standard deviation by itself fails to accurately distinguish between the sections.

31. For greater accuracy and understanding, we use 3 decimal places and avoid shortcut formulas.

a. the original population

x	x-μ	(x-μ)^2
1	-1	1
2	0	0
3	1	1
6	0	2

$\mu = (\Sigma x)/N = 6/3 = 2$

$\sigma^2 = \Sigma(x-\mu)^2/N = 2/3 = .667$
$\sigma = .816$

b. the nine samples: using $s^2 = \Sigma(x-\bar{x})^2/(n-1)$ [for each sample, n = 2]

sample	$\bar{x}$	s^2	s
1,1	1.0	0	0
1,2	1.5	0.5	0.707
1,3	2.0	2.0	1.414
2,1	1.5	0.5	0.707
2,2	2.0	0	0
2,3	2.5	0.5	0.707
3,1	2.0	2.0	1.414
3,2	2.5	0.5	0.707
3,3	3.0	0	0
	18.0	6.0	5.656

mean of the 9 calculated variances

$(\Sigma s^2)/9 = 6.0/9 = 2/3 = .667$

c. the nine samples: using $\sigma^2 = \Sigma(x-\mu)^2/N$ [for each sample, N = 2]

sample	μ	σ^2
1,1	1.0	0
1,2	1.5	0.25
1,3	2.0	1.00
2,1	1.5	0.25
2,2	2.0	0
2,3	2.5	0.25
3,1	2.0	1.00
3,2	2.5	0.25
3,3	3.0	0
	18.0	3.00

mean of the 9 calculated variances

$(\Sigma \sigma^2)/9 = 3.0/9 = 1/3 = .333$

d. The approach in (b) of dividing by n-1 when calculating the sample variance gives a better estimate of the population variance. On the average, the approach in (b) gave the correct population variance of $2/3 = .667$. The approach in (c) of dividing by n underestimated the correct population variance. When computing sample variances, divide by n-1 and not by n.

e. No. An unbiased estimator is one that gives the correct answer on the average. Since the average value of s^2 in part (b) was .667, which was the correct value calculated for σ^2 in part (a), s^2 is an unbiased estimator of σ^2. Since the average value of s in part (b) is $(\Sigma s)/9 = 5.656/9 = .628$, which is not the correct value of .816 calculated for σ in part (a), s is not an unbiased estimator of σ.

NOTE: Since the average value of $\overline{x}$ in part (b) is $(\Sigma \overline{x})/9 = 18.0/9 = 2.0$, which is the correct value calculated for μ in part (a), $\overline{x}$ is an unbiased estimator of μ.

2-6 Measures of Position

1. a. $x - \mu = 130 - 100 = 30$
   b. $30/\sigma = 30/15 = 2.00$
   c. $z = (x - \mu)/\sigma = (130 - 100)/15 = 30/15 = 2.00$
   d. The z score in part (c) and the number of standard deviations by which a score differs from the mean in part (b) are identical, being different names for the same concept.

3. In general, $z = (x - \mu)/\sigma$.
   a. $z_{85} = (85 - 69.0)/2.8 = 5.71$
   b. $z_{64} = (64 - 69.0)/2.8 = -1.79$
   c. $z_{69.72} = (69.72 - 69.0)/2.8 = 0.26$

5. $z = (x - \mu)/\sigma$
   $z_{70} = (70 - 63.6)/2.5 = 2.56$
   Yes, that height is considered unusual since $2.56 > 2.00$.

7. $z = (x - \mu)/\sigma$
   $z_{101} = (101 - 98.20)/0.62 = 4.52$
   Yes, that temperature is unusually high, since $4.52 > 2.00$. It suggests that either the person is healthy but has a very unusual temperature for a healthy person, or that person is sick (i.e., has an elevated temperature attributable to some cause).

9. In general $z = (x - \overline{x})/s$
   history: $z_{75} = (75 - 80)/12 = -0.42$
   psychology: $z_{27} = (27 - 30)/8 = -0.375$
   The psychology score has the better relative position since $-0.375 > -.042$

11. preliminary values: $n = 36$, $\Sigma x = 29.4056$, $\Sigma x^2 = 24.02111984$
    $\overline{x} = (\Sigma x)/n = 29.4056/36 = .81682$
    $s^2 = [n(\Sigma x^2) - (\Sigma x)^2]/[n(n-1)]$
    $\quad = [36(24.02111984) - (29.4056)^2]/[36(35)]$
    $\quad = .07100288/1260$
    $\quad = .000056351$
    $s = .007507$
    $z = (x - \overline{x})/s$
    $z_{.7901} = (.7901 - .81682)/.007507 = -3.56$
    Yes; since $-3.56 < -2.00$, .7901 is an unusual weight for regular Coke.

13. Let b = # of scores below x; n = total number of scores.
    In general, the percentile of score x is (b/n)·100.
    The percentile for a weight of .8264 is (33/36)·100 = 92.

15. Let b = # of scores below x; n = total number of scores.
    In general, the percentile of score x is (b/n)·100.
    The percentile for a weight of .8192 is (22/36)·100 = 61.

17. To find P_{80}, L = (80/100)·36 = 28.8 rounded up to 29.
    Since the 29th score is .8229, P_{80} = .8229.

19. To find D_6 = P_{60}, L = (60/100)·36 = 21.6 rounded up to 22.
    Since the 22nd score is .8189, D_6 = .8189.

21. To find Q_3 = P_{75}, L = (75/100)·36 = 27 -- a whole number.
    The mean of the 27th and 28th scores, Q_3 = (.8207 + .8211)/2 = .8209.

23. To find D_1 = P_{10}, L = (10/100)·36 = 3.6 rounded up to 4.
    Since the 4th score is .8073, D_1 = .8073.

NOTE: For exercises 25-36, refer to the ordered cross-numbered chart at the right. A cross-numbered chart gives position in the list. The column head gives the tens digit and the row lead gives the ones digit, so that **270** is #43 in the ordered list of the 54 weights.

	0	1	2	3	4	5
0		65	120	166	220	365
1	26	76	125	166	236	416
2	29	79	132	180	262	436
3	34	80	140	182	**270**	446
4	40	86	140	202	316	514
5	45	90	144	202	332	
6	48	94	148	204	344	
7	60	105	150	204	348	
8	62	114	150	212	356	
9	64	116	154	220	360	

25. Let b = # of scores below x; n = total number of scores
    In general, the percentile of score x is (b/n)·100.
    The percentile of score 144 is (24/54)·100 = 44.

27. Let b = # of scores below x; n = total number of scores
    In general, the percentile of score x is (b/n)·100.
    The percentile of score 316 is (43/54)·100 = 80.

29. To find P_{85}, L = (85/100)·54 = 45.9 rounded up to 46.
    Since the 46th score is 344, P_{85} = 344.

31. To find Q_1 = P_{25}, L = (25/100)·54 = 13.5 rounded up to 14.
    Since the 14th score is 86, Q_1 = 86.

33. To find D_9 = P_{90}, L = (90/100)·54 = 48.6 rounded up to 49.
    Since the 49th score is 360, D_9 = 360.

35. To find $\tilde{x}$ = P_{50}, L = (50/100)·54 = 27 -- a whole number.
    The mean of the 27th and 28th scores, P_{50} = (150 + 150)/2 = 150.0.

37. In general, z scores are not affected by the particular unit of measurement that is used. The relative position of a score (whether it is above or below the mean, its rank in an ordered list of the scores, etc.) is not affected by the unit of measurement, and relative position is what a z score communicates. Mathematically, the same units (feet, centimeters, dollars, etc.) appear in both the numerator and denominator of $(x-\mu)/\sigma$ and cancel out to leave the z score unit-free. In fact the z score is also called the *standard score* for that very reason –

it is a standardized value that is independent of the unit of measure employed.

NOTE: In more technical language, changing from one unit of measure to another (feet to centimeters, °F to °C, dollars to pesos, etc.) is a linear transformation -- i.e., if x is the score in one unit, then (for some appropriate values of a and b) $y = ax + b$ is the score in the other unit. In such cases it can be shown (see exercises 2.5 #30) that

$$\mu_y = a\mu_x + b \text{ and } \sigma_y = a\sigma_x.$$

The new z score is the same the old one, since

$$z_y = (y - \mu_y)/\sigma_y$$
$$= [(ax + b) - (a\mu_x + b)]/a\sigma_x$$
$$= [ax - a\mu_x]/a\sigma_x$$
$$= (x - \mu_x)/\sigma_x$$
$$= z_x$$

2-7 Exploratory Data Analysis

The exercises in this section may be done much more easily when ordered lists of the values are available. On the next page appear ordered lists for the data of the odd-numbered exercises. The left-most column gives the ordered ID numbers 1-50. Data sets with more than 50 values have more than one column – the second column being ordered values 51-100, etc.

NOTE: A boxplot can be misleading if an extreme value artificially extends the whisker beyond the reasonable data. This manual follows the convention of extending the whisker only to the highest and lowest values within $1.5(Q_3 - Q_1)$ of Q_2 – any point more than 1.5 times the width of the box away from the median will be considered an outlier and represented separately. This type of boxplot is discussed in exercise #11. The boxplots given in this manual can be converted to the ones presented in the text by extending the whisker to the most extreme outlier.

1. Consider the 25 employee ages.
   For $Q_1 = P_{25}$, $L = (25/100) \cdot 25 = 6.25$ rounded up to 7.
   For $\tilde{x} = Q_2 = P_{50}$, $L = (50/100) \cdot 25 = 12.5$ rounded up to 13.
   For $Q_3 = P_{75}$, $L = (75/100) \cdot 25 = 18.75$ rounded up to 19.

$\min = x_1 = 34$
$Q_1 = x_7 = 65$
$Q_2 = x_{13} = 75$
$Q_3 = x_{19} = 84$
$\max = x_{25} = 88$

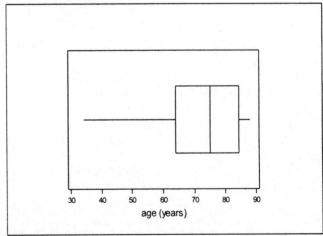

The employees seem to be considerably older than most American workers.

The following ordered lists are used for the odd-numbered exercises.

	#1 age	#3 pul	#3 pul	#5 can	#5 can	#5 can	#5 can	#7 mal	#7 fem	#9 author	#9 umass
01	34	8	71	205	277	287	301	31	21	28.35	18.95
02	39	15	72	210	277	287	301	32	24	33.35	24.00
03	41	40	72	210	279	287	302	32	26	35.00	26.70
04	45	47	72	211	279	287	302	32	26	38.45	27.95
05	53	48	72	215	279	288	303	33	26	39.15	28.00
06	63	52	72	216	279	288	303	35	17	41.95	34.95
07	65	52	72	222	279	288	303	36	28	42.65	38.70
08	72	54	73	225	279	288	303	37	30	44.00	41.60
09	73	55	75	227	279	288	304	37	30	44.65	41.95
10	73	55	75	230	280	288	304	38	31	50.15	45.60
11	74	58	75	231	280	289	305	39	31	50.65	48.70
12	75	60	76	243	280	289	305	39	33	51.35	50.70
13	75	60	76	244	281	289	306	40	33	53.35	52.00
14	76	60	77	246	281	290	306	40	34	53.35	54.70
15	77	60	77	247	281	290	306	41	34	54.35	57.35
16	79	60	78	247	281	290	307	42	34	57.55	60.00
17	79	60	78	247	281	290	308	42	34	58.65	60.00
18	82	60	78	250	281	290	309	43	35	59.35	67.00
19	84	60	78	253	282	291	310	43	35	60.00	67.70
20	85	61	78	255	282	291	311	44	35	60.00	70.00
21	86	63	78	255	282	291	313	45	37	61.00	72.70
22	86	63	80	256	282	291	314	45	37	61.00	73.00
23	87	63	80	257	283	292	315	46	38	61.00	75.35
24	87	63	80	260	283	292	317	47	39	62.00	76.00
25	88	63	80	260	283	292	504	48	41	63.60	77.00
26		64	80	262	283	292		48	41	64.55	78.00
27		64	80	262	283	293		51	41	65.65	79.20
28		64	80	262	283	293		53	42	69.35	80.55
29		65	81	263	284	293		55	44	75.00	82.00
30		65	82	265	284	294		56	49	75.65	83.35
31		66	83	266	284	294		56	50	75.95	84.30
32		66	83	268	284	295		60	60	77.80	86.00
33		67	84	268	284	295		60	61	80.00	86.70
34		67	85	269	284	295		61	61	80.00	88.00
35		67	86	269	284	295		62	74	88.00	89.35
36		67	88	270	284	296		76	80		92.00
37		67	88	270	285	296					92.70
38		68	88	271	285	296					94.00
39		68	90	272	285	297					95.00
40		68	90	272	285	297					102.95
41		69	92	273	285	297					
42		69	92	274	285	298					
43		69	97	274	285	298					
44		69	100	275	286	298					
45		70		275	286	299					
46		70		275	286	299					
47		70		275	286	300					
48		70		276	287	300					
49		70		276	287	300					
50		71		276	287	300					

3. Consider the 94 pulse rates.

For $Q_1 = P_{25}$, $L = (25/100) \cdot 94 = 23.5$ rounded up to 24.

For $\tilde{x} = Q_2 = P_{50}$, $L = (50/100) \cdot 94 = 47$ -- an integer, use 47.5.

For $Q_3 = P_{75}$, $L = (75/100) \cdot 94 = 70.5$ rounded up to 71.

min $= x_1 = 8$

$Q_1 = x_{24} = 63$

$Q_2 = x_{47.5} = (70 + 70)/2 = 70$

$Q_3 = x_{71} = 78$

max $= x_{94} = 100$

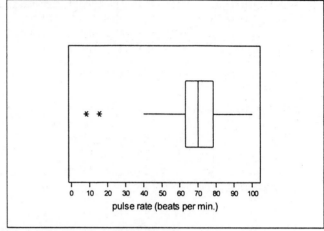

The values 8 and 15 are far from the other values and appear to be errors.

NOTE for exercise #3 and all future uses of this data set: The values 8 and 15 are obvious errors that will be eliminated from all subsequent analyses. Such obvious errors occur in many real life data sets. Often correct values can be deduced and the data adjusted accordingly. If, for example, all the other values were multiples of 4, one could infer that students monitored their pulse for 15 seconds and multiplied by 4 to obtain a per minute rate -- and that some students forgot to multiply by 4. It appears that the instructor gathered data by having each student monitor his own pulse rate -- and that there was not careful instruction or a re-take of suspicious values. It also appears that 6 students couldn't find their pulse -- and those places were just left blank. Although also questionable, the values in the 40's will be included in all subsequent analyses of the pulse data, but the numbers are suspect and should not be taken as accurate renderings of student pulse rates.

5. Consider the 175 axial loads of the 0.0111 in. thick cans.

For $Q_1 = P_{25}$, $L = (25/100) \cdot 175 = 43.75$ rounded up to 44.

For $\tilde{x} = Q_2 = P_{50}$, $L = (50/100) \cdot 175 + 87.5$ rounded up to 88.

For $Q_3 = P_{75}$, $L = (75/100) \cdot 175 = 131.25$ rounded up to 132.

min $= x_1 = 205$

$Q_1 = x_{44} = 275$

$Q_2 = x_{88} = 285$

$Q_3 = x_{132} = 295$

max $= x_{175} = 504$

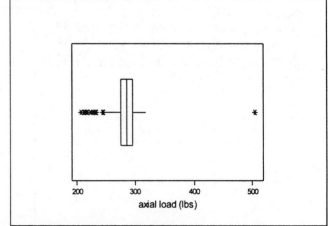

The value 504 appears to be either an error or an anomaly.

7. Consider the 36 actor and 36 actress values.
   For $Q_1 = P_{25}$, $L = (25/100) \cdot 36 = 9$ -- an integer, use 9.5.
   For $\tilde{x} = Q_2 = P_{50}$, $L = (50/100) \cdot 36 = 18$ -- an integer, use 18.5.
   For $Q_3 = P_{75}$, $L = (75/100) \cdot 36 = 27$ -- an integer, use 27.5.

   For the actors
   min $= x_1 = 31$
   $Q_1 = x_{9.5} = (37 + 38)/2 = 37.5$
   $Q_2 = x_{18.5} = (43 + 43)/2 = 43$
   $Q_3 = x_{27.5} = (51 + 53)/2 = 52$
   max $= x_{36} = 76$

   For the actresses
   min $= x_1 = 21$
   $Q_1 = x_{9.5} = (30 + 31)/2 = 30.5$
   $Q_2 = x_{18.5} = (35 + 35)/2 = 35$
   $Q_3 = x_{27.5} = (41 + 42)/2 = 41.5$
   max $= x_{36} = 80$

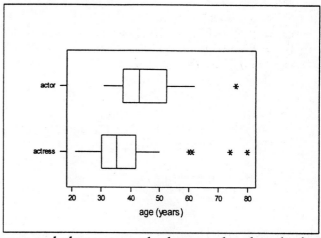

The ages for the actresses cover a wider range and cluster around a lower value than do the ages of the actors.

9. Consider the 35 new textbook prices for the author's college.
   For $Q_1 = P_{25}$, $L = (25/100) \cdot 35 = 8.75$ rounded up to 9.
   For $\tilde{x} = Q_2 = P_{50}$, $L = (50/100) \cdot 35 = 17.5$ rounded up to 18.
   For $Q_3 = P_{75}$, $L = (75/100) \cdot 35 = 26.25$ rounded up to 27.

   min $= x_1 = 28.35$
   $Q_1 = x_9 = 44.65$
   $Q_2 = x_{18} = 59.35$
   $Q_3 = x_{27} = 65.65$
   max $= x_{34} = 88.00$

   Consider the 40 new textbook prices for UMASS.
   For $Q_1 = P_{25}$, $L = (25/100) \cdot 40 = 10$ -- an integer, use 10.5.
   For $\tilde{x} = Q_2 = P_{50}$, $L = (50/100) \cdot 40 = 20$ -- an integer, use 20.5.
   For $Q_3 = P_{75}$, $L = (75/100) \cdot 40 = 30$ -- an integer, use 30.5.

   min $= x_1 = 18.95$
   $Q_1 = x_{10.5} = (45.60 + 48.70)/2 = 47.15$
   $Q_2 = x_{20.5} = (70.00 + 72.70)/2 = 71.35$
   $Q_3 = x_{30.5} = (83.35 + 84.30)/2 = 83.825$
   max $= x_{40} = 102.95$

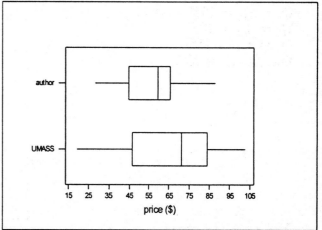

New textbooks at the University of Massachusetts appear to cost more than those at the author's school.

11. The following given values are also calculated in detail in exercise #5.
    $Q_1 = 275 \quad Q_2 = 285 \quad Q_3 = 295$
    a. IQR = $Q_3 - Q_1$ = 295 - 275 = 20
    b. Since 1.5(IQR) = 1.5(20) = 30, the modified boxplot line extends to
       the smallest value above Q_1-30 = 275-30 = 245, which is 246
       the largest value below Q_3+30 = 295+30 = 325, which is 317
    c. Since 3.0(IQR) = 3.0(30) = 60, mild outliers are x values for which
       $\quad Q_1\text{-}60 \le x < Q_1\text{-}30 \quad$ or $\quad Q_3\text{+}30 < x \le Q_3\text{+}60$
       $\quad\quad 215 \le x < 245 \quad\quad\quad\quad\quad 325 < x \le 355$
       Those values are as follows.
       lower end: 215,216,222,225,227,230,231,243,244
       upper end: (none)
    d. Extreme outliers in this exercise are x values for which x < 215 or x > 355.
       Those values are as follows.
       lower end: 205,210,210,211
       upper end: 504

Review Exercises

1. The scores arranged in order are:
   42 43 46 46 47 48 49 49 50 51 51 51 51 51 52 52 54 54 54 54 55
   55 55 55 56 56 56 57 57 57 57 58 60 61 61 61 62 64 64 65 68 69
   preliminary values: n = 42, Σx = 2304, Σx^2 = 128,014
   a. $\bar{x} = (\Sigma x)/n = (2304)/42 = 54.9$
   b. $\tilde{x} = (55 + 55)/2 = 55.0$
   c. M = 51
   d. m.r. = (42 + 69)/2 = 55.5
   e. R = 69 - 42 = 27
   f. s = 6.3 (from part g)
   g. $s^2 = [n(\Sigma x^2) - (\Sigma x)^2]/[n(n-1)]$
      = $[42(128,014) - (2304)^2]/[42(41)]$
      = (68,172)/1722
      = 39.6
   h. For $Q_1 = P_{25}$, L = (25/100)·42 = 10.5 rounded up to 11.
      And so $Q_1 = x_{11} = 51$.
   i. For P_{30}, L = (30/100)·42 = 12.6 rounded up to 13.
      And so $P_{30} = x_{13} = 51$.
   j. For $D_7 = P_{70}$, L = (70/100)·42 = 29.4 rounded up to 30.
      And so $D_7 = x_{30} = 57$.

2. a. $z = (x - \bar{x})/s$
      $z_{42} = (42 - 54.857)/6.292$
      = -2.04
   b. Yes; Teddy Roosevelt's inaugural age is unusual since -2.04 < -2.00.
   c. According to the Range Rule of Thumb, the usual values are within 2s of $\bar{x}$
      usual minimum: $\bar{x}$ - 2s = 54.9 - 2(6.3) = 42.3
      usual maximum: $\bar{x}$ + 2s = 54.9 + 2(6.3) = 67.5
   In addition to the 42 in part (b), the ages 68 an 69 are also unusual.

3.

age	frequency
40 – 44	2
45 – 49	6
50 – 54	12
55 – 59	12
60 – 64	7
65 – 69	3
	42

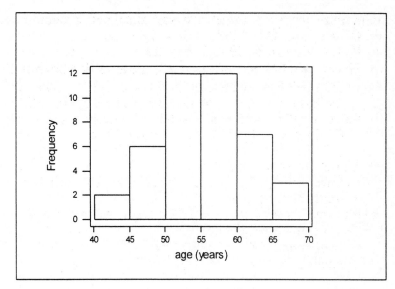

4. See the figure above at the right.

NOTE: Unlike other data, age is usually not reported to the nearest unit. The bars in a histogram extend from class boundary to class boundary. Because of the way that ages are reported, the boundaries here are 40, 45, 50, etc -- i.e., someone 44.9 years old still reports an age of 44 and crosses into the next class only upon turning 45, not upon turning 44.5.

5. The scores are given in order in exercise #1.

For $Q_1 = P_{25}$, L = $(25/100) \cdot 42 = 10.5$, rounded up to 11.

For $\tilde{x} = Q_2 = P_{50}$, L = $(50/100) \cdot 42 = 21$ -- an integer, use 21.5.

For $Q_3 = P_{75}$, L = $(75/100) \cdot 42 = 31.5$, rounded up to 32.

The 5-number summary is:

min = x_1 = 42

$Q_1 = x_{11} = 51$

$Q_2 = x_{21.5} = (55 + 55)/2 = 55$

$Q_3 = x_{32} = 58$

max = $x_{42} = 69$

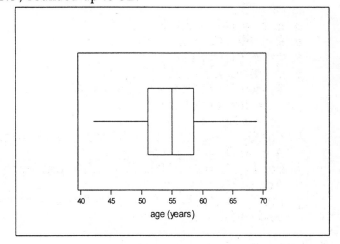

6. a. Since scores 48.6 to 61.2 are within 1·s of $\overline{x}$, the *Empirical Rule for Data with a Bell-Shaped Distribution* states that about 68% of such persons fall within those limits.

b. Since scores 42.3 to 67.5 are within 2·s of $\overline{x}$, the *Empirical Rule for Data with a Bell-Shaped Distribution* states that about 95% of such persons fall within those limits.

7. In general, $z = (x - \overline{x})/s$.

management, $z_{72} = (72 - 80)/12 = -0.67$

production, $z_{19} = (19 - 20)/5 = -0.20$

The score on the test for production employees has the better relative position since $-0.20 > -0.67$.

8. In general, the *Range Rule of Thumb* states that the range typically covers about 4 standard deviations -- with the lowest and highest scores being about 2 standard deviations below and above the mean respectively. The following answers assume the ordered textbook ages range from x_1 = 0 years (i.e., copyrighted during the current year) to x_n = 12 years.
   a. The estimated mean age is $(x_1 + x_n)/2 = (0 + 12)/2$
$$= 6 \text{ years.}$$
   b. The estimated standard deviation of the ages is $R/4 = (x_n - x_1)/4$
$$= (12 - 0)/4$$
$$= 3 \text{ years.}$$

9. In general, adding the same value to every score will move the all scores up the number line by the same amount. This changes the location of the scores on the number line, but it does not affect the spread of the scores or the shape of the distribution -- i.e., each measure of center will change by the amount added, but the measures of variation will not be affected.
   a. Since adding 5.0 minutes to every score moves all the scores 5 units up on the number line, the mean will increase by 5.0 to 12.3 + 5.0 = 17.3 minutes.
   b. Since adding 5.0 minutes to every score does not affect the spread of the scores, the standard deviation will not change and remain 4.0 minutes.
   c. Since adding 5.0 minutes to every score does not affect the spread of the scores, the variance will not change and will remain $(4.0)^2 = 16.0$.

10. Arranging the categories in decreasing order by frequency produces the following figure.

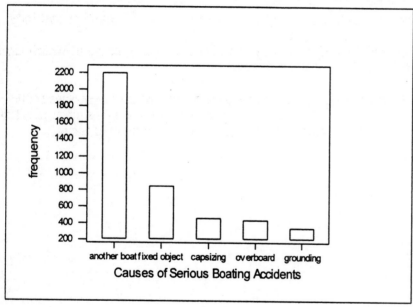

Cumulative Review Exercises

1. The scores arranged in order are: 0 1 4 7 7 7 11 12 12 12 13 18 19 23 25 30
   preliminary values: n = 16, Σx = 201, Σx^2 = 3625
   a. $\bar{x}$ = (Σx)/n = (201)/16 = 12.6
      $\tilde{x}$ = (12 + 12)/2 = 12.0
      M = 7, 12 (bi-modal)
      m.r. = (0 + 30)/2 = 15.0
   b. R = 30 - 0 = 30
      s^2 = [n(Σx^2) - (Σx)^2]/[n(n-1)] = [16(3625) - (201)^2]/[16(15)] = (17599)/240 = 73.3
      s = 8.6
   c. continuous. Even though the values were reported rounded to whole years, age can be
      any non-negative value on a continuum.
      NOTE: While age is continuous, it could be that the population from which the sample
      was selected was the ages rounded to whole years -- and so it may be argued that "the
      population from which the given years were drawn" was discrete.
   d. ratio. Differences are consistent and there is a meaningful zero; 8 years is twice as much
      time as 4 years.

2. a. mode. The median requires at least ordinal level data, and the mean and the midrange
      require at least interval level data.
   b. convenience. The group was not selected by any process other than the fact that they
      happened to be the first names on the list.
   c. cluster. The population was divided into units (election precincts), some of which were
      selected at random to be examined in their entirety.

3. No; using the mean of the state values counts each state equally, while states with more
   people will have a greater affect on the per capita consumption for the population in all 50
   states combined. Use the state populations as weights to find the weighted mean.

Chapter 3

Probability

3-2 Fundamentals

1. Since $0 \leq P(A) \leq 1$ is always true, the following values less than 0 or greater than 1 cannot be probabilities.
   values less than 0: -1
   values greater than 1: 2   5/3 $\sqrt{2}$

3. Let H = getting a head when a balanced coin is tossed.
   H is one of 2 equally likely different outcomes (head or tail); use Rule 2.
   P(H) = 1/2

5. a. There are 144 + 1354 = 1498 M&M candies in the package.
      Let O = getting an orange M&M.
      O includes 144 of 1498 equally likely outcomes; use Rule 2.
      P(O) = 144/1498 = .0961
   b. There are 21 + 8 + 26 + 33 + 5 + 7 = 100 M&M candies in the data set.
      Let O = getting an orange M&M.
      O includes 8 of 100 equally likely outcomes; use Rule 2.
      P(O) = 8/100 = .08
   c. Yes; approximately, when (a) and (b) are considered individually. But considered together, the results from parts (a) and (b) suggest that the proportion of M&M's that are orange may be slightly less than 10%.

7. Let H = selecting an at bat for which McGwire hit a home run.
   H includes 70 of 509 equally likely outcomes; use Rule 2.
   P(H) = 70/509 = .138

9. The sample space of the 36 outcomes for tossing a pair of dice is given below.
   ```
   1-1  2-1  3-1  4-1  5-1  6-1
   1-2  2-2  3-2  4-2  5-2  6-2
   1-3  2-3  3-3  4-3  5-3  6-3
   1-4  2-4  3-4  4-4  5-4  6-4
   1-5  2-5  3-5  4-5  5-5  6-5
   1-6  2-6  3-6  4-6  5-6  6-6
   ```
   NOTE: Don't be confused by the fact that the dice look identical. Like identical twins, they are still different entities -- call them die A and die B. Outcome 1-6 (getting a 1 on die A and a 6 on die B), for example, is different from 6-1 (getting a 6 on die A and a 1 on die B).
   a. Let T = getting a total of 2
      T includes only 1 (the 1-1) of 36 equally likely outcomes; use Rule 2.
      P(T) = 1/36 = .028
   b. Yes; since P(T) < .05, T is considered an unusual event.

11. a. Let C = answering a question correctly with a random guess.
       C includes only 1 of 10 equally likely outcomes; use Rule 2.
       P(C) = 1/10 = .1
    b. No; since P(C) > .05, C is not considered an unusual event.

13. a. Let L = an American Airline flight arrives late.
    L occurred 400 - 344 = 56 times in a sample of 400; use Rule 1.
    P(L) ≈ 56/400 = .14
    b. No; since P(L) > .05, L is not considered an unusual event.
    NOTE: On time arrival rate by itself does not describe an airline's record. By lengthening the projected flight times, the airline could achieve a higher on time arrival rate but would have planes creating problems by arriving "too early" for their unrealistic (and inaccurate) scheduled arrival times.

15. NOTE: This problem can be approached two ways. Finding P(l of your 2 numbers matches the winning number), the usual approach, requires using the Addition Rule of section 3-3. Finding P(the winning number matches one of your 2 numbers), which is an equivalent event, requires only the methods of this section.
    a. The winning number can be selected in 25,827,165 equally likely ways.
    Let W = you winning the grand prize
    W can occur 2 ways, if either of your numbers is selected; use Rule 2.
    P(W) = 2/25,827,165 = .0000000774
    b. Yes; since P(W) < .05, W is an unusual event.

17. There were 255 + 2302 older Americans surveyed.
    Let A = an older American has Alzheimer's disease
    A occurred 255 times in a sample of 2302; use Rule 1.
    P(A) ≈ 255/2557 = .0997
    Yes; since about 1 in 10 older Americans is affected, it is a major concern.

19. There were 225 + 275 = 500 donors surveyed.
    Let O = a person has type O blood.
    A occurred 225 times in a sample of 500; use Rule 1.
    P(O) ≈ 225/500 = .45

21. a. Let B = a person's birthday is October 18.
    B is one of 365 (assumed) equally likely outcomes; use Rule 2.
    P(B) = 1/365 = .00274
    b. Let O = a person's birthday is in October.
    O includes 31 of 365 (assumed) equally likely outcomes; use Rule 2.
    P(N) = 31/365 = .0849

23. There were 132 + 880 = 1012 respondents.
    Let D = getting a respondent who gave the doorstop answer.
    D includes 132 of 1012 equally likely outcomes; use Rule 2.
    P(D) = 132/1012 = .130

25. Let A = selecting a driver in that age bracket has an accident.
    A occurred 136 times in a sample of 400; use Rule 1.
    P(A) ≈ 136/400 = .340
    Yes; since this indicates more than 1 such driver in every 3 is likely to have an accident, this should be a concern.
    NOTE: This is properly an estimate that such a person <u>had</u> an accident last year, and not that such a person <u>will have</u> an accident next year. To extend the estimate as desired requires the additional assumption that relevant factors (speed limits, weather, alcohol laws, the economy, etc.) remain the same.

27. Let V = a Viagra user experiences a headache.
    V occurred 117 times in a sample of 117 + 617 = 734; use Rule 1.
    P(V) ≈ 117/734 = .159

29. a. The 4 equally likely outcomes are: BB BG GB GG.
    b. P(2 girls) = 1/4 = .25
    c. P(1 of each sex) = 2/4 = .5

31. a. Let M = Mean wins the race. It is given that P(M) = 3/17.
    $P(\overline{M})$ = 1 - P(M) = 1 - 3/17 = 14/17
    odds against Mean winning = $P(\overline{M})$/P(M) = (14/17)/(3/17) = 14/3 = 14 to 3
    b. The text's formula *(payoff odds) = (net profit)/(amount bet)* may be solved for
    (net profit) to yield (net profit) = (payoff odds)·(amount bet) = (4/1)·($4) = $16.

33. Chance fluctuations in testing situations typically prevent subjects from scoring exactly the
    same each time the test is given. When treated subjects show an improvement, there are
    two possibilities: (1) the treatment was effective and the improvement was due to the
    treatment or (2) the treatment was not effective and the improvement was due to chance.
    The accepted standard is that an event whose probability of occurrence is less than .05 is
    an unusual event. In this scenario, the probability that the treatment group shows an
    improvement even if the drug has no effect is calculated to be .04. If the treatment group
    does show improvement, then, either the treatment truly is effective or else an unusual
    event has occurred. Operating according to the accepted standard, one should conclude
    that the treatment was effective.

35. If the odds against A are a:b, then P(A) = b/(a+b).
    Let M = Millennium wins the next race.
    If the odds against M are 3:5, then a=3 and b=5 in the statement above.
    P(M) = 5/(3+5) = 5/8 = .625

3-3 Addition Rule

1. a. No, it's possible for MTV to be the principal news source of a voter under the age of
      30.
   b. No, it's possible for someone treated with an experimental drug to experience improved
      symptoms.
   c. Yes, if the outcome is an even number then it cannot be an odd number.

3. a. $P(\overline{A})$ = 1 - P(A)
         = 1 - .05
         = .95
   b. P(girl) = $P(\overline{boy})$
             = 1 - P(boy)
             = 1 - .513
             = .487

5. Make a chart like the one on the right.
   let F = getting a female
       G = getting a person coded green

		COLOR		
		Red	Green	
GENDER	Male	3	2	5
	Female	4	6	10
		7	8	15

   There are two approaches.
   * Use broad categories and allow for double-counting (i.e., the "formal addition rule").
   P(F or G) = P(F) + P(G) - P(F and G)
             = 10/15 + 8/15 - 6/15
             = 12/15 = .8
   * Use individual mutually exclusive categories that involve no double-counting (i.e., the
   "intuitive addition rule"). For simplicity in this problem, we use MR for "M and R" and
   FR for "F and R" and so on.

P(F or G) = P(FR or MG or FG)
           = P(FR) + P(MG) + P(FG)
           = 4/15 + 2/15 + 6/15
           = 12/15 = .8

NOTE: In general, using broad categories and allowing for double-counting is a "more powerful" technique that "lets the formula do the work" and requires less analysis by the solver. Except when such detailed analysis is instructive, this manual uses the first approach.

NOTE: Throughout the manual we follow the pattern in exercise #5 of using the first letter [or other natural designation] of each category to represent that category. And so in exercise #7 P(D) = P(selecting a person who died) and P(M) = P(selecting a man) and so on. If there is ever cause for ambiguity, the notation will be clearly defined. Since mathematics and statistics use considerable notation and formulas, it is important to clearly define what various letters stand for.

7. Make a chart like the one on the right.
   a. P(W or G) = P(W) + P(G) - P(W and G)
               = 422/2223 + 45/2223 - 0/2223
               = 467/2223
               = .210
   b. P(W or S) = P(W) + P(S) - P(W and S)
               = 422/2223 + 706/2223 - 318/2223
               = 810/2223
               = .364

	FATE		
	Survived	Died	
Men	332	1360	1692
GROUP Women	318	104	422
Boys	29	35	64
Girls	27	18	45
	706	1517	2223

9. Let B = a person's birthday is October 18
   $P(\bar{B})$ = 1 - P(B)
        = 1 - 1/365
        = 364/365 = .997

11. Make a chart like the one on the right.
    P(M or Y) = P(M) + P(Y) - P(M and Y)
              = 50/82 + 47/82 - 39/82
              = 58/82 = .707

		TICKET?		
		Yes	No	
GENDER	male	39	11	50
	female	8	24	32
		47	35	82

13. Make a chart like the one on the right.
    $P(\bar{A})$ = 1 - P(A)
         = 1 - 40/100
         = 60/100 = .60

		Rh FACTOR		
		+	−	
	A	35	5	40
GROUP	B	8	2	10
	AB	4	1	5
	O	39	6	45
		86	14	100

15. Refer to exercise #13.
    P(A or Rh-) = P(A) + P(Rh-) - P(A and Rh-)
                = 40/100 + 14/100 - 5/100
                = 49/100 = .49

17. Refer to exercise #13.
    $P(\overline{Rh+})$ = 1 - P(Rh+)
              = 1 - 86/100
              = 14/100 = .14

19. Refer to exercise #13.
    P(AB or Rh+) = P(AB) + P(Rh+) - P(AB and Rh+)
                 = 5/100 + 86/100 - 4/100
                 = 87/100 = .87

21. Make a chart like the one on the right.
    Let A = a person is age 18-21
        N = a person does not respond
    P(A or N) = P(A) + P(N) - P(A and N)
              = 84/359 + 31/359 - 11/359
              = 104/359 = .290

		RESPOND?		
		Yes	No	
AGE	18-21	73	11	84
	22-29	255	20	275
		328	31	359

23. Make a chart like the one on the right.
    P(N) = 580/1000 = .580

		HARASSMENT?		
		Yes	No	
GENDER	male	240	380	620
	female	180	200	380

25. The general formula is: P(A or B) = P(A) + P(B) - P(A and B).
    Solving for P(A and B) yields: P(A and B) = P(A) + P(B) - P(A or B)
    a. For P(A) = 3/11 and P(B) = 4/11 and P(A or B) = 7/11, the Addition Rule produces
       P(A and B) = P(A) + P(B) - P(A or B)
                  = 3/11 + 4/11 - 7/11
                  = 0
    This means that A and B are mutually exclusive.
    b. For P(A) = 5/18 and P(B) = 11/18 and P(A or B) = 13/18, the Addition Rule
       produces
       P(A and B) = P(A) + P(B) - P(A or B)
                  = 5/18 + 11/18 - 13/18
                  = 3/18
    This means that A and B are not mutually exclusive.

27. If the *exclusive or* is used instead of the *inclusive or*, then the double-counted probability
    must be completely removed (i.e., must be subtracted twice) and the formula becomes
    P(A or B) = P(A) + P(B) - 2·P(A and B)

3-4 Multiplication Rule: Basics

1. a. Independent, since getting a head on the first flip does not affect the outcome on the
      second flip.
   b. Dependent, since speeding while driving to class makes it more likely to get a traffic
      ticket while driving to class.
   c. Independent, since the car's not starting does not affect the performance of the kitchen
      light.
      NOTE: One could argue that a car's not starting may be due to improper maintenance --
      the same type of behavior that would be likely, for example, to leave burned out kitchen
      light bulbs unchanged. If so, then a person's car not starting suggests he may be one
       who doesn't properly maintain things -- which suggests that his kitchen light might not
      be working either. This manual ignores such highly speculative hypothetical scenarios.

3. There are 5 red + 4 yellow + 3 green = 12 lenses.
   let A = the first lens selected is yellow
       B = the second lens selected is yellow
   In general, P(A and B) = P(A)·P(B|A)
   a. P(A and B) = P(A)·P(B|A) = (4/12)·(4/12)
                              = 1/9 = .111
   b. P(A and B) = P(A)·P(B|A) = (4/12)·(3/11)
                              = 1/11 = .091

5. let F = selecting a female
   P(F) = 15/25, for the first selection only
   $P(F_1 \text{ and } F_2) = P(F_1) \cdot P(F_2 | F_1)$
   $= (15/25) \cdot (14/24)$
   $= 7/20 = .35$

7. let W = selecting a wrong answer
   P(W) = 4/5, for each selection
   $P(W_1 \text{ and } W_2 \text{ and } W_3) = P(W_1) \cdot P(W_2) \cdot P(W_3) = (4/5) \cdot (4/5) \cdot (4/5) = 64/125 = .512$

9. a. let J = a person is born on July 4
      P(J) = 1/365, for each selection
      $P(J_1 \text{ and } J_2) = P(J_1) \cdot P(J_2)$
      $= (1/365) \cdot (1/365)$
      $= 1/133225 = .00000751$

   b. let S = a person's birthday is favorable to being the same
      P(S) = 365/365 for the first person, but then the second person must match
      P(the birthdays are the same) $= P(S_1 \text{ and } S_2)$
      $= P(S_1) \cdot P(S_2 | S_1)$
      $= (365/365) \cdot (1/365)$
      $= 1/365 = .00274$

11. Since the sample size is no more than 5% of the size of the population, treat the selections as being independent even though the selections are made without replacement.
    let G = a selected CD is good
    P(G) = .97, for each selection
    P(batch accepted) = P(all good)
    $= P(G_1 \text{ and } G_2 \text{ and}...\text{and } G_{12})$
    $= P(G_1) \cdot P(G_2) \cdot ... \cdot P(G_{12})$
    $= (.97)^{12} = .694$

13. let G = a girl is born
    P(G) = .50, for each couple
    $P(G_1 \text{ and } G_2 \text{ and}...\text{and } G_{10}) = P(G_1) \cdot P(G_2) \cdot ... \cdot P(G_{10})$
    $= (.50)^{10} = .000977$

    Yes, the gender selection method appears to be effective. The choice is between two possibilities: (1) the gender selection method has no effect and a very unusual event occurred or (2) the gender selection method is effective.

15. Since the sample size is no more than 5% of the size of the population, treat the selections as being independent even though the selections are made without replacement.
    let L = a person is left-handed
    P(L) = .10, for each selection
    a. $P(L_1 \text{ and } L_2 \text{ and } L_3) = P(L_1) \cdot P(L_2) \cdot P(L_3)$
       $= (.10)(.10)(.10)$
       $= .001$
    b. $P(L_1 \text{ and } L_2 \text{ and}...\text{and } L_{20}) = P(L_1) \cdot P(L_2) \cdot ... \cdot P(L_{20})$
       $= (.10)^{20}$
       $= .00000000000000000001$
    The probability of having 20 persons randomly chosen from the general population all be left-handed is so small that it is essentially not possible. Something is wrong. Either the students are not from the general population (e.g., are part of a program to help left-handers cope in a right-handed world) or (more likely) the survey is flawed because the respondents either are not telling the truth or miscoded their responses due to unclear directions.

17. let T = the tire named is consistent with all naming the same tire
    P(T) = 4/4 for the first person, but then the others must match him
    P(all name the same tire) = P(T_1 and T_2 and T_3 and T_4)
    $\qquad\qquad\qquad\qquad\qquad$ = P(T_1)·P(T_2|T_1)·P(T_3|T_1 and T_2)·P(T_4|T_1 and T_2 and T_3)
    $\qquad\qquad\qquad\qquad\qquad$ = (4/4)(1/4)(1/4)(1/4)
    $\qquad\qquad\qquad\qquad\qquad$ = 1/64 = .0156

19. Since the sample size is no more than 5% of the size of the population, treat the selections
    as being independent even though the selections are made without replacement.
    let N = a cell phone is not defective
    P(N) = 1 - .04 = .96, for each selection
    P(N_1 and N_2 and...and N_{25}) = P(N_1)·P(N_2)·...·P(N_{25})
    $\qquad\qquad\qquad\qquad\qquad\qquad$ = (.96)^{25}
    $\qquad\qquad\qquad\qquad\qquad\qquad$ = .360

    No, there is not strong evidence to conclude that the new process is better. Even if the
    new process made no difference, the observed result is not considered unusual and could
    be expected to occur about 36% of the time.

21. Refer to the table at the right.
    Since the sample size is no more than
      5% of the size of the population,
      treat the selections as being inde-
      pendent even though the selections
      are made without replacement.
    P(Pos) = 1054/20000
    $\qquad$ = .0527, for each selection
    P(Pos_1 and Pos_2) = P(Pos_1)·P(Pos_2)
    $\qquad\qquad\qquad\qquad$ = (.0527)·(.0527)
    $\qquad\qquad\qquad\qquad$ = .00278

		TEST RESULT		
		Pos	Neg	
HIV?	Yes	57	3	60
	No	997	18,943	19,940
		1054	18,946	20,000

23. Refer to the table and comments for exercise #21.
    P(Yes) = 60/20000 = .003, for each selection
    P(Yes_1 and Yes_2) = P(Yes_1)·P(Yes_2)
    $\qquad\qquad\qquad\qquad$ = (.003)·(.003)
    $\qquad\qquad\qquad\qquad$ = .000009

25. let D = a birthday is different from any yet selected
    P(D_1) = 366/366   NOTE: With nothing to match, it <u>must</u> be different.
    P(D_2|D_1) = 365/366
    P(D_3|D_1 and D_2) = 364/366
    ...
    P(D_5|D_1 and D_2 and...and D_4) = 362/366

    P(D_{25}|D_1 and D_2 and...and D_{24}) = 342/366
    a. P(all different) = P(D_1 and D_2 and D_3)
    $\qquad\qquad\qquad\qquad$ = P(D_1)·P(D_2)·P(D_3)
    $\qquad\qquad\qquad\qquad$ = (366/366)·(365/366)·(364/366)
    $\qquad\qquad\qquad\qquad$ = .992
    b. P(all different) = P(D_1 and D_2 and...and D_5)
    $\qquad\qquad\qquad\qquad$ = P(D_1)·P(D_2)·...·P(D_5)
    $\qquad\qquad\qquad\qquad$ = (366/366)·(365/366)·...·(362/366)
    $\qquad\qquad\qquad\qquad$ = .973
    c. P(all different) = P(D_1 and D_2 and...and D_{25})
    $\qquad\qquad\qquad\qquad$ = P(D_1)·P(D_2)·...·P(D_{25})
    $\qquad\qquad\qquad\qquad$ = (366/366)·(365/366)·...·(342/366)
    $\qquad\qquad\qquad\qquad$ = .432

NOTE: A program to perform this calculation can be constructed using a programming language, or using most spreadsheet or statistical software packages.
In BASIC, for example, use

```
10 LET P=1
15 PRINT "How many birthdays?"
20 INPUT D
30   FOR K=1 TO D-1
40   LET P=P*(366-K)/366
50   NEXT K
55 PRINT "The probability they all are different is"
60 PRINT P
70 END
```

27. This is problem can be done by two different methods. In either case,
    let A = getting an ace
        S = getting a spade
    * consider the sample space
    The first card could be any of 52 cards; for each first card, there are 51 possible second cards. This makes a total of $52 \cdot 51 = 2652$ equally likely outcomes in the sample space. How many of them are A_1S_2?
    The aces of hearts, diamonds and clubs can be paired with any of the 13 clubs for a total of $3 \cdot 13 = 39$ favorable possibilities. The ace of spades can only be paired with any of the remaining 12 members of that suit for a total of 12 favorable possibilities. Since there are $39 + 12 = 51$ favorable possibilities among the equally likely outcomes,
    $$P(A_1S_2) = 51/2652$$
    $$= .0192$$
    * use the formulas
    Let As and Ao represent the ace of spades and the ace of any other suit respectively. Break A_1S_2 into mutually exclusive parts so the probability can be found by adding and without having to consider double-counting.
    $$P(A_1S_2) = P[(As_1 \text{ and } S_2) \text{ or } (Ao_1 \text{ and } S_2)]$$
    $$= P(As_1 \text{ and } S_2) + P(Ao_1 \text{ and } S_2)$$
    $$= P(As_1) \cdot P(S_2|As_1) + P(Ao_1) \cdot P(S_2|Ao_1)$$
    $$= (1/52)(12/51) + (3/52)(13/51)$$
    $$= 12/2652 + 39/2652$$
    $$= 51/2652$$
    $$= .0192$$

3-5 Multiplication Rule: Complements and Conditional Probability

1. If it is not true that "at least one of them is a girl," then "they are all boys."

3. If it is not true that "all students receive a grade of A," then "at least one student does not receive a grade of A."

5. let B = a child is a boy
   $P(B) = .5$, for each birth
   $$P(\text{at least one girl}) = 1 - P(\text{all boys})$$
   $$= 1 - P(B_1 \text{ and } B_2 \text{ and } B_3)$$
   $$= 1 - P(B_1) \cdot P(B_2) \cdot P(B_3)$$
   $$= 1 - (.5) \cdot (.5) \cdot (.5)$$
   $$= 1 - .125$$
   $$= .875$$

7. let N = rolling a number that is not a six
P(N) = 5/6, for each roll
P(at least one six) = 1 - P(all numbers that are not six)
$$= 1 - P(N_1 \text{ and } N_2 \text{ and } N_3 \text{ and } N_4 \text{ and } N_5)$$
$$= 1 - P(N_1) \cdot P(N_2) \cdot P(N_3) \cdot P(N_4) \cdot P(N_5)$$
$$= 1 - (5/6) \cdot (5/6) \cdot (5/6) \cdot (5/6) \cdot (5/6)$$
$$= 1 - 3125/7776$$
$$= 4651/7776 = .598$$

9. let H = getting a head
P(H) = .5, for each toss
$$P(H_3 \mid H_1 \text{ and } H_2) = P(H_3) = .5$$

11. let B = a person has given birth to a baby
W = a person is a woman
$$P(W \mid B) = 1 \quad [\text{a certainty}]$$

13. let F = the alarm clock fails
P(F) = .01, for each clock
P(at least one works) = 1 - P(all fail)
$$= 1 - P(F_1 \text{ and } F_2 \text{ and } F_3)$$
$$= 1 - P(F_1) \cdot P(F_2) \cdot P(F_3)$$
$$= 1 - (.01)(.01)(.01)$$
$$= 1 - .000001$$
$$= .999999$$

NOTE: Rounded to 3 significant digits as usual, the answer is 1.00. In cases when rounding to 3 significant digits produces a probability of 1.00, this manual gives the answer with sufficient significant digits to distinguish the answer from a certainty.

15. let N = a person is HIV negative
P(N) = .9, for each person in the at-risk population
P(HIV positive result) = P(at least person is HIV positive)
$$= 1 - P(\text{all persons HIV negative})$$
$$= 1 - P(N_1 \text{ and } N_2 \text{ and } N_3)$$
$$= 1 - P(N_1) \cdot P(N_2) \cdot P(N_3)$$
$$= 1 - (.9)(.9)(.9)$$
$$= 1 - .729$$
$$= .271$$

NOTE: This plan is very efficient. Suppose, for example, there were 3,000 people to be tested. Only in .271 = 27.1% of the groups would a retest need to be done for each of the 3 individuals. Those (.271)·(1,000) = 271 groups would generate 271·3 = 813 retests. The total number of tests required is then 1813 (1000 original tests + 813 retests), only 60% of the 3,000 tests that would have been required to test everyone individually.

17. Refer to the table at the right.
This problem may be done two ways.
* reading directly from the table
$$P(M \mid D) = 1360/1517$$
$$= .897$$
* using the formula
$$P(M \mid D) = P(M \text{ and } D)/P(D)$$
$$= (1360/2223)/(1517/2223)$$
$$= .6118/.6824$$
$$= .897$$

		FATE		
		Survived	Died	
	Men	332	1360	1692
GROUP	Women	318	104	422
	Boys	29	35	64
	Girls	27	18	45
		706	1517	2223

NOTE: In general, the manual will use the most obvious approach -- which most often is the first one of applying the basic definition by reading directly from the table.

19. Refer to the table and comments for exercise #17.
    P([B or G]$|$S) = [29 + 27]/706
    $\qquad$ = 56/706 = .0793

21. For 25 randomly selected people,
    a. P(no 2 share the same birthday) = .432 [see exercise #25c in section 3-4]
    b. P(at least 2 share the same birthday) = 1 - P(no 2 share the same birthday)
    $\qquad\qquad\qquad\qquad\qquad\qquad\qquad\qquad$ = 1 - .432
    $\qquad\qquad\qquad\qquad\qquad\qquad\qquad\qquad$ = .568

3-6 Counting

1. 14! = 14·13·12·11·10·9·8·7·6·5·4·3·2·1 = 87,178,291,200
   NOTE: It's staggering how such a small number like 14 can have such a large value for its factorial. In practice this means that if you have 14 books on a shelf you could arrange them in over 87 billion different orders!

3. $_{50}P_3$ = 50!/47! = (50·49·48·47!)/47! = 50·49·48 = 117,600
   NOTE: This technique of "cancelling out" or "reducing" the problem by removing the factors 47! = 47·96·...·1 from both the numerator and the denominator is preferred over actually evaluating 50!, actually evaluating 47!, and then dividing those two very large numbers. In general, a smaller factorial in the denominator can be completely divided into a larger factorial in the numerator to leave only the "excess" factors not appearing the in the denominator. This is the technique employed in this manual -- e.g., see #5 below, where the 43! is cancelled from both the numerator and the denominator. In addition, $_nP_r$ and $_nC_r$ will always be integers; calculating a non-integer value for either expression indicates an error has been made. More generally, the answer to any <u>counting</u> problem (but not a <u>probability</u> problem) must always be a whole number; a fractional number indicates that an error has been made.

5. Let W = winning the described lottery with a single selection
   The total number of possible combinations is
   $\quad _{49}C_6$ = 49!/(43!6!)
   $\qquad\quad$ = (49·48·47·46·45·44)/6!
   $\qquad\quad$ = 13,983,816
   Since only one combination wins, P(W) = 1/13,983,816

7. Let W = winning the described lottery with a single selection
   The total number of possible combinations is
   $\quad _{44}C_6$ = 44!/(38!6!)
   $\qquad\quad$ = (44·43·42·41·40·39)/6!
   $\qquad\quad$ = 7,059,052
   Since only one combination wins, P(W) = 1/7,059,052

9. let W = winning the described lottery with a single selection
   The total number of possible arrangements is
   $\quad _{51}P_6$ = 51!/45!
   $\qquad\quad$ = 51·50·49·48·47·46
   $\qquad\quad$ = 12,966,811,200
   Since only one arrangement wins, P(W) = 1/12,966,811,200

11. a. Since a different arrangement of selected names is a different slate, use
    $$_{12}P_4 = 12!/8!$$
    $$= 12 \cdot 11 \cdot 10 \cdot 9 = 11,880$$
    b. Since a different arrangement of selected names is the same committee, use
    $$_{12}C_4 = 12!/(8!4!)$$
    $$= (12 \cdot 11 \cdot 10 \cdot 9)/4! = 495$$

13. There are 5 players. Once the 3 are chosen for the United Way event, the other 2 for the Heart Fund event are automatically determined. The problem reduces to how many ways can 3 be selected for the United Way Event. Since a different arrangement of the selected names is the same delegation, use
    $$_5C_3 = 5!/(2!3!)$$
    $$= (5 \cdot 4)/2! = 10$$

15. a. $4! = 4 \cdot 3 \cdot 2 \cdot 1 = 24$
    b. let A = the cities are in alphabetical order
    Since only one of the routes has the cities in alphabetical order, $P(A) = 1/24$

17. let S = the resulting route is one with the shortest distance
    The total number of routes is $8! = 40,320$
    Since 2 of the routes represent the shortest distance, $P(S) = 2/40,320 = 1/20,160$

19. The first note is given by *. There are 3 possibilities (R,U,D) for each of the next 15 notes. There are $3 \cdot 3 \cdot 3 \cdot 3 \cdot 3 \cdot 3 \cdot 3 \cdot 3 \cdot 3 \cdot 3 \cdot 3 \cdot 3 \cdot 3 \cdot 3 \cdot 3 = 3^{15}$
    $$= 14,348,907 \text{ possible sequences.}$$
    NOTE: This assumes each song has at least 16 notes.

21. a. There are 2 possibilities (B,G) for each baby. The number of possible sequences is
    $$2 \cdot 2 \cdot 2 \cdot 2 \cdot 2 \cdot 2 \cdot 2 \cdot 2 \cdot 2 \cdot 2 \cdot 2 \cdot 2 \cdot 2 \cdot 2 \cdot 2 \cdot 2 \cdot 2 \cdot 2 \cdot 2 \cdot 2 = 2^{20}$$
    $$= 1,048,576$$
    b. The number of ways to arrange 20 items consisting of 10 B's and 10 G's is
    $$20!/(10!10!) = 20 \cdot 19 \cdot 18 \cdot 17 \cdot 16 \cdot 15 \cdot 14 \cdot 13 \cdot 12 \cdot 11/10!$$
    $$= 184,756$$
    c. Let E = getting equal numbers of boys and girls in 20 births.
    Based on parts (a) and (b), $P(E) = 184756/1048576$
    $$= .176.$$
    d. The preceding results indicate that getting 10 boys and 10 girls when 20 babies are randomly selected is not an unusual event -- for one such random selection. But the probability of getting that result twice in a row would be $(.176) \cdot (.176) = .0310$, which would be considered unusual since $.0310 < .05$. Getting 10 boys and 10 girls "consistently" should arouse suspicion.

23. Use the factorial rule with n = 12 to determine the total number of possible arrangements.
    $$n! = 479,001,600$$
    Let I = selecting the houses so that the readings are in increasing order.
    Since only one of the equally likely possibilities gives the readings in increasing order,
    $$P(I) = 1/479,001,600 = .00000000209$$
    Yes, her concern about the test equipment is justified. NOTE: The houses were presumably measured in a geographical order and not randomly. If she is working her way toward the center of a radon concentration, such a pattern would be expected. Since radon dissipates in circulating air (home readings are typically taken in basements), an extra measurement in a well-ventilated upstairs room should determine whether there is an equipment malfunction.

25. The total number of ways of selecting 5 eggs from among 12 is

$_{12}C_5 = 12!/(7!5!)$
$= (12 \cdot 11 \cdot 10 \cdot 9 \cdot 8)/5! = 792.$

a. Let A = selecting all 3 of the cracked eggs. The total number of ways of selecting 3 eggs from among the 3 cracked ones and 2 eggs from among the 9 good ones is

$_3C_3 \cdot _9C_2 = [3!/(0!3!)] \cdot [9!/(7!2!)]$
$= [1] \cdot [36] = 36$
$P(A) = 36/792 = .0455$

b. Let N = selecting none of the cracked eggs. The total number of ways of selecting 0 eggs from among the 3 cracked ones and 5 eggs from among the 9 good ones is

$_3C_0 \cdot _9C_5 = [3!/(3!0!)] \cdot [9!/(4!5!)]$
$= [1] \cdot [126] = 126$
$P(N) = 126/792 = .1591$

c. Let T = selecting 2 of the cracked eggs. The total number of ways of selecting 2 eggs from among the 3 cracked ones and 3 eggs from among the 9 good ones is

$_3C_2 \cdot _9C_3 = [3!/(1!2!)] \cdot [9!/(6!3!)]$
$= [3] \cdot [84] = 252$
$P(T) = 252/792 = .3182$

NOTE 1: Since the answer in part (a) required 4 decimal places to provide 3 significant digit accuracy, the answers in parts (b) and (c) are also given to 4 decimal places for consistency and ease of comparison. In general, this pattern will apply throughout the manual.

NOTE 2: There is only one possibility not covered in parts (a)-(c) above -- viz., one of the cracked eggs is selected. As a check, it is recommended that this probability be calculated. If these 4 mutually exclusive and exhaustive probabilities do not sum to 1.000, then an error has been made.

Let O = selecting 1 of the cracked eggs.
The total number of ways of selecting 1 egg from among the 3 cracked ones and 4 eggs from among the 9 good ones is $_3C_1 \cdot _9C_4 = [3!/(2!1!)] \cdot [9!/(5!4!)]$
$= [3] \cdot [126] = 378$

$P(O) = 378/792 = .4773$
$P(N \text{ or } O \text{ or } T \text{ or } A) = P(N) + P(O) + P(T) + P(A)$
$= .1591 + .4773 + .3182 + .0455$
$= 1.000 \text{ [the 4 values actually sum to 1.0001 due to round-off]}$

And so everything seems in order. Performing an extra calculation that can provide a check to both the methodology and the mathematics is a good habit that is well worth the effort.

27. This problem may be approached in two different ways.

* general probability formulas of earlier sections
  Let F = selecting a card favorable for getting a flush.
  $P(F_1) = 52/52$, since the first card could be anything
  $P(F_2) = 12/51$, since the second card must be from the same suit as the first
  etc.
  $P(\text{getting a flush}) = P(F_1 \text{ and } F_2 \text{ and } F_3 \text{ and } F_4 \text{ and } F_5)$
  $= P(F_1) \cdot P(F_2) \cdot P(F_3) \cdot P(F_4) \cdot P(F_5)$
  $= (52/52) \cdot (12/51) \cdot (11/50) \cdot (10/49) \cdot (9/48) = .00198$

* counting techniques of this section
  The total number of possible 5 card selections is
  $_{52}C_5 = 52!/(47!5!) = 2,598,960$
  The total number of possible 5 card selections from one particular suit is
  $_{13}C_5 = 13!/(8!5!) = 1287$
  The total number of possible 5 card selections from any of the 4 suits is
  $4 \cdot 1287 = 5148$
  $P(\text{getting a flush}) = 5148/2598960 = .00198$

29. There are 26 possible first characters, and 36 possible characters for the other positions. Find the number of possible names using 1,2,3,...,8 characters and then add to get the total.

characters	possible names		
1	26	=	26
2	26·36	=	936
3	26·36·36	=	33,696
4	26·36·36·36	=	1,213,056
5	26·36·36·36·36	=	43,670,016
6	26·36·36·36·36·36	=	1,572,120,576
7	26·36·36·36·36·36·36	=	56,596,340,736
8	26·36·36·36·36·36·36·36	=	2,037,468,266,496
	total	=	2,095,681,645,538

Review Exercises

1. Refer to the table at the right.
   let Y = a subject experienced soreness
   N = a subject did not experience soreness
   D = a subject used the drug Nicorette
   P = a subject used the placebo
   P(D) = 152/305 = .498

		TREATMENT		
		Drug	Placebo	
SORENESS?	Yes	43	35	78
	No	109	118	227
		152	153	305

2. Refer to the table and notation of exercise #1.
   P(N) = 227/305 = .744

3. Refer to the table and notation of exercise #1.
   P(D or Y) = P(D) + P(Y) - P(D and Y)
   $\qquad$ = 152/305 + 78/305 - 43/305
   $\qquad$ = 187/305 = .613

4. Refer to the table and notation of exercise #1.
   P(P or N) = P(P) + P(N) - P(P and N)
   $\qquad$ = 153/305 + 227/305 - 118/305
   $\qquad$ = 262/305 = .859

5. Refer to the table and notation of exercise #1.
   $P(Y_1 \text{ and } Y_2) = P(Y_1) \cdot P(Y_2 | Y_1)$
   $\qquad$ = (78/305)·(77/304) = .0648

6. Refer to the table and notation of exercise #1.
   $P(P_1 \text{ and } P_2) = P(P_1) \cdot P(P_2 | P_1)$
   $\qquad$ = (153/305)·(152/304) = .251

7. Refer to the table and notation of exercise #1.
   P(Y | P) = 35/153 = .229

8. Refer to the table and notation of exercise #1.
   P(Y or P) = P(Y) + P(P) - P(Y and P)
   $\qquad$ = 78/305 + 153/305 - 35/305
   $\qquad$ = 196/305 = .643

9. let G = an item is defective
   P(G) = .27, for each item
   a. $P(\overline{G}) = 1 - .27 = .73$
   b. $P(G_1 \text{ and } G_2) = P(G_1) \cdot P(G_2)$
      $= (.27) \cdot (.27) = .0729$
   c. P(at least one good) = 1 - P(all not good)
      $= 1 - P(\overline{G}_1 \text{ and } \overline{G}_2 \text{ and } \overline{G}_3)$
      $= 1 - P(\overline{G}_1) \cdot P(\overline{G}_2) \cdot P(\overline{G}_3)$
      $= 1 - (.73)^3 = .611$

10. Since the sample size is no more than 5% of the size of the population, treat the selections
    as being independent even though the selections are made without replacement.
    let G = a CD is good
    P(G) = .98, for each selection
    P(reject batch) = P(at least one is defective)
      = 1 - P(all are good)
      $= 1 - P(G_1 \text{ and } G_2 \text{ and } G_3 \text{ and } G_4)$
      $= 1 - P(G_1) \cdot P(G_2) \cdot P(G_3) \cdot P(G_4)$
      $= 1 - (.98)^4$
      = .0776
    NOTE: P(reject batch) = 1 - (2450/2500)·(2449/2499)·(2448/2498)·(2447/2497) = .0777
    if the problem is done considering the effect of the non-replacement.

11. let B = getting a boy
    P(B) = .5, for each child
    P(at least one girl) = 1 - P(all boys)
      $= 1 - P(B_1 \text{ and } B_2 \text{ and } B_3 \text{ and } B_4 \text{ and } B_5)$
      $= 1 - P(B_1) \cdot P(B_2) \cdot P(B_3) \cdot P(B_4) \cdot P(B_5)$
      $= 1 - (.5)^5$
      = .969

12. a. let W = the committee of 3 is the 3 wealthiest members
       The number of possible committees of 3 is $_{10}C_3 = 10!/(7!3!) = 120$.
       Since only one of those committees is the 3 wealthiest members,
       P(W) = 1/120 = .00833
    b. Since order is important, the number of possible slates of officers is
       $_{10}P_3 = 10!/7! = 720$

13. let E = a winning even number occurs
    a. P(E) = 18/38 [=9/19]
    b. odds against E = $P(\overline{E})/P(E)$
       = (20/38)/(18/38)
       = 20/18 = 10/9, usually expressed as 10:9
    c. payoff odds = (net profit):(amount bet)
       If the payoff odds are 1:1, the net profit is $5 for a winning $5 bet.

14. Since the sample size is no more than 5% of the size of the population, treat the selections
    as being independent even though the selections are made without replacement.
    let R = getting a Republican
    P(R) = .30, for each selection
    $P(R_1 \text{ and } R_2 \text{ and}...\text{and } R_{12}) = P(R_1) \cdot P(R_2) \cdot ... \cdot P(R_{12}) = (.30)^{12}$
      = .000000531
    No, the pollster's claim is not correct.

15. let L = a selected 27 year old male lives for one more year
    P(L) = .9982, for each such person
    $P(L_1 \text{ and } L_2 \text{ and}...\text{and } L_{12}) = P(L_1) \cdot P(L_2) \cdot ... \cdot P(L_{12}) = (.9982)^{12}$
    $= .979$

16. let G = a selected switch is good
    P(G) = 44/48, for the first selection only
    a. $P(G_1 \text{ and } G_2 \text{ and } G_3 \text{ and } G_4) = P(G_1) \cdot P(G_2) \cdot P(G_3) \cdot P(G_4)$
        $= (44/48) \cdot (43/47) \cdot (42/46) \cdot (41/45)$
        $= 3258024/4669920$
        $= .69766$
    Since this is the only way to arrange 4G's,
        P(all 4 are good) = .698
    b. $P(G_1 \text{ and } G_2 \text{ and } G_3 \text{ and } \overline{G}_4) = P(G_1) \cdot P(G_2) \cdot P(G_3) \cdot P(\overline{G}_4)$
        $= (44/48) \cdot (43/47) \cdot (42/46) \cdot (4/45)$
        $= 317856/4669920$
        $= .06806$
    Since 3G's and $1\overline{G}$ can be arranged in 4!/(3!1!) = 4 ways (of which this is only one),
        P(exactly 3 are good) = 4·(.06806) = .272
    c. $P(G_1 \text{ and } G_2 \text{ and } \overline{G}_3 \text{ and } \overline{G}_4) = P(G_1) \cdot P(G_2) \cdot P(\overline{G}_3) \cdot P(\overline{G}_4)$
        $= (44/48) \cdot (43/47) \cdot (4/46) \cdot (3/45)$
        $= 22704/4669920$
        $= .00486$
    Since 2G's and $2\overline{G}$'s can be arranged in 4!/(2!2!) = 6 ways (of which this is only one),
        P(exactly 2 are good) = 6·(.00486)
            $= .0292$

Cumulative Review Exercises

1. scores arranged in order: 2 4 7 8 10 10 13 13 14 15 17 29
    summary statistics: n = 12, $\Sigma x = 142$, $\Sigma x^2 = 2222$
    a. $\overline{x} = (\Sigma x)/n = 142/12 = 11.8$
    b. $\tilde{x} = (10 + 13)/2 = 11.5$
    c. $s^2 = [n \cdot (\Sigma x^2) - (\Sigma x)^2]/[n(n-1)]$
        $= [12(2222) - (142)^2]/[(12)(11)]$
        $= 49.24$
        s = 7.0
    d. $s^2 = 49.2$ [from part (c)]
    e. minimum usual age = $\overline{x} - 2s = 11.8 - 2(7.0) = -2.2$, truncated to 0
       maximum usual age = $\overline{x} + 2s = 11.8 + 2(7.0) = 25.8$
       No, a new jet [i.e., x = 0] is within the above limits and would not be considered unusual.
    f. ratio, since differences are meaningful and there is a natural meaning for 0
    g. let C = getting the correct answer
       Since there are 4 given choices and only 1 of them is correct,
           P(C) = 1/4
    h. let A = getting an age of at least 10
       Since there are 12 ages and 8 of them are at least 10,
           P(A) = 8/12 = .667
    i. let L = getting an age less than 10
       P(L) = 4/12, for the first selection only
           $P(L_1 \text{ and } L_2) = P(L_1) \cdot P(L_2 | L_1) = (4/12) \cdot (3/11) = .0909$

   j. let L = getting an age less than 10
      P(L) = 4/12, for each selection
         $P(L_1 \text{ and } L_2) = P(L_1) \cdot P(L_2) = (4/12) \cdot (4/12) = .1111$

2. Let x = the height selected.
   The values identified on the boxplot are: x_1 [minimum] = 56.1
                                               P_{25} = 62.2
                                               P_{50} = 63.6
                                               P_{75} = 65.0
                                               x_n [maximum] = 71.1
   NOTE: Assume that the number of heights is so large that (1) repeated sampling does not affect the probabilities for subsequent selections and (2) it can be said that P_a has a% of the heights below it and (100-a)% of the heights above it.
   a. The exact values of $\bar{x}$ and s cannot be determined from the information given. The Range Rule of Thumb [the highest and lowest values are approximately 2 standard deviations above and below the mean], however, indicates that $s \approx (x_n - x_1)/4$ and $\bar{x} \approx (x_1 + x_n)/2$. In this case we estimate $\bar{x} \approx (56.1 + 71.1)/2 = 63.6$
      NOTE: The boxplot, which gives $\tilde{x} = P_{50} = 63.6$, is symmetric. This suggests that the original distribution is approximately so. In a symmetric distribution the mean and median are equal. This also suggests $\bar{x} \approx 63.6$.
   b. $P(56.1 < x < 62.2) = P(x_1 < x < P_{25})$
                        $= .25 - 0$
                        $= .25$
   c. $P(x < 62.2 \text{ or } x > 63.6) = P(x < 62.2) + P(x > 63.6)$
                              $= P(x < P_{25}) + P(x > P_{50})$
                              $= .25 + .50$
                              $= .75$

      NOTE: No height is less than 62.2 <u>and</u> greater than 63.6 -- i.e., the addition rule for mutually exclusive events can be used.
   d. Let B = selecting a height between 62.2 and 63.6.
      $P(B) = P(P_{25} < x < P_{50})$
           $= .50 - .25$
           $= .25$
      $P(B_1 \text{ and } B_2) = P(B_1) \cdot P(B_2 | B_1)$
                     $= (.25) \cdot (.25)$
                     $= .0625$
   e. let S = a selected woman is shorter than the mean
         T = a selected woman is taller than the mean
         E = the event that a group of 5 women consists of 3 T's and 2 S's
      The total number of arrangements of T's and S's is $2 \cdot 2 \cdot 2 \cdot 2 \cdot 2 = 32$
      The number of arrangements of 3 T's and 2 S's is $5!/(3!2!) = 10$
       P(E) = 10/32 = .3125
      NOTE: The exact value of the mean cannot be determined from the information given. From part (a) we estimate $\bar{x} \approx 63.6$, which is also the median. If this is true then *in this particular problem* $P(x < \bar{x}) = P(S) = .50$ and $P(x > \bar{x}) = P(T) = .50$. This means that all arrangements of T's and S's are equally likely and
       P(E) = (# of ways E can occur)/(total number of possible outcomes)

Chapter 4

Probability Distributions

4-2 Random Variables

1. a. Continuous, since weight can be any value on a continuum.
   b. Discrete, since the cost must be an integer number cents.
   c. Continuous, since time can be any value on a continuum.
   d. Continuous, since volume can be any value on a continuum.
   e. Discrete, since the number of cans must be an integer.

NOTE: When working with probability distributions and formulas in the exercises that follow, always keep these important facts in mind.
 * If one of the conditions for a probability distribution does not hold, the formulas do not apply and produce numbers that have no meaning.
 * $\Sigma x \cdot P(x)$ gives the mean of the x values and must be a number between the highest and lowest x values.
 * $\Sigma x^2 \cdot P(x)$ gives the mean of the x^2 values and must be a number between the highest and lowest x^2 values.
 * $\Sigma P(x)$ must always equal 1.000.
 * Σx and Σx^2 have no meaning and should not be calculated.
 * The quantity $[\Sigma x^2 \cdot P(x) - \mu^2]$ cannot possibly be negative; if it is, then there is a mistake.
 * Always be careful to use the <u>unrounded</u> mean in the calculation of the variance and to take the square root of the <u>unrounded</u> variance to find the standard deviation.

3. This is a probability distribution since $\Sigma P(x)=1$ is true and $0 \le P(x) \le 1$ is true for each x.

x	P(x)	x·P(x)	x²	x²P(x)	
0	.125	0	0	0	$\mu = \Sigma x \cdot P(x)$
1	.375	.375	1	.375	$= 1.500$, rounded to 1.5
2	.375	.750	4	1.500	$\sigma^2 = \Sigma x^2 \cdot P(x) - \mu^2$
3	.125	.375	9	1.125	$= 3.000 - (1.500)^2$
	1.000	1.500		3.000	$= .750$
	1.000	.601		.875	$\sigma = .866$, rounded to 0.9

5. This is a probability distribution since $\Sigma P(x)=1$ is true and $0 \le P(x) \le 1$ is true for each x.

x	P(x)	x·P(x)	x²	x²P(x)	
0	.0625	0	0	0	$\mu = \Sigma x \cdot P(x)$
1	.2500	.2500	1	.2500	$= 2.0000$, rounded to 2.0
2	.3750	.7500	4	1.5000	$\sigma^2 = \Sigma x^2 \cdot P(x) - \mu^2$
3	.2500	.7500	9	2.2500	$= 5.0000 - (2.0000)^2$
4	.0625	.2500	16	1.0000	$= 1.0000$
	1.0000	2.0000		5.0000	$\sigma = 1.0000$, rounded to 1.0

7. This is a probability distribution since $\Sigma P(x)=1$ is true and $0 \le P(x) \le 1$ is true for each x.

x	P(x)	x·P(x)	x²	x²P(x)	
0	.512	0	0	0	$\mu = \Sigma x \cdot P(x)$
1	.301	.301	1	.301	$= .730$, rounded to 0.7
2	.132	.264	4	.528	$\sigma^2 = \Sigma x^2 \cdot P(x) - \mu^2$
3	.055	.165	9	.495	$= 1.324 - (.73)^2$
	1.000	.730		1.324	$= .791$
					$\sigma = .889$, rounded to 0.9

9. This is a probability distribution since $\Sigma P(x)=1$ is true and $0 \le P(x) \le 1$ is true for each x.

x	P(x)	x·P(x)	x²	x²P(x)	
0	.36	0	0	0	$\mu = \Sigma x \cdot P(x)$
1	.48	.48	1	.48	$= .80$, rounded to 0.8
2	.16	.32	4	.64	$\sigma^2 = \Sigma x^2 \cdot P(x) - \mu^2$
	1.00	.80		1.12	$= 1.12 - (.80)^2$
					$= .48$
					$\sigma = .693$, rounded to 0.7

11. This is a probability distribution since $\Sigma P(x)=1$ is true and $0 \le P(x) \le 1$ is true for each x.

x	P(x)	x·P(x)	x²	x²P(x)	
0	.289	0	0	0	$\mu = \Sigma x \cdot P(x)$
1	.407	.407	1	.407	$= 1.101$, rounded to 1.1
2	.229	.458	4	.916	$\sigma^2 = \Sigma x^2 \cdot P(x) - \mu^2$
3	.065	.195	9	.585	$= 2.077 - (1.101)^2$
4	.009	.036	16	.144	$= .865$
5	.001	.005	25	.025	$\sigma = .939$, rounded to 0.9
	1.000	1.101		2.077	

13. a.

x	P(x)	x·P(x)	
175	1/38	175/38	$E = \Sigma x \cdot P(x)$
-5	37/38	-185/38	$= -10/38$
	38/38	-10/38	$= \$-0.263$ [i.e., a loss of 26.3¢]

The expected loss is 26.3¢ for a $5 bet.

b.

x	P(x)	x·P(x)	
5	18/38	90/38	$E = \Sigma x \cdot P(x)$
-5	20/38	-100/38	$= -10/38$
	38/38	-10/38	$= \$-0.263$ [i.e., a loss of 26.3¢]

The expected loss is 26.3¢ for a $5 bet.

c. Since the bets in parts (a) and (b) each have (exactly the same) negative expectation, the best option is not to bet at all.

NOTE: Problems whose probabilities do not "come out even" as decimals can be very sensitive to rounding errors. Since the P(x) values are used in several subsequent calculations, express them as exact fractions instead of rounded decimals. If it is not convenient to continue with fractions throughout the entire problem, then use sufficient decimal places (typically one morethan "usual") in the x·P(x) and x²·P(x) columns to guard against cumulative rounding errors.

15. a. Mike "wins" $100,000 - $250 = $99,750 if he dies.
 Mike "loses" $250 if he lives.

b.

x	P(x)	x·P(x)	
99750	.0015	149.625	$E = \Sigma x \cdot P(x)$
-250	.9985	-249.625	$= -100.000$ (i.e., a loss of $100)
	1.0000	-100.000	

c. Since the company is making a $100 profit at this price, it would break even if it sold the policy for $100 less -- i.e. for $150. NOTE: This oversimplified analysis ignores the cost of doing business. If the costs (printing, salaries, etc.) associated with offering the policy is $25, for example, then the company's profit is only $75 and selling the policy for $75 less (i.e., for $175) would represent the break even point for the company.

17. Use the *Range Rule of Thumb* to identify unusual results.
 The minimum usual value is $\mu - 2\sigma = 11.4 - 2 \cdot (.75) = 9.9$.
 Yes; since "fewer than 8" is less than 9.9, that would be considered an unusual result.

19. a. Since the probabilities must sum to 1.000, the missing value (found by subtracting all the other values from 1.000) is P(x=2) = .161.

b. Use probabilities to identify unusual results.

P(at least 3 sevens) = P(x≥3)
= P(x=3 or x=4 or x=5)
= P(x=3) + P(x=4) + P(x=5)
= .032 + .003 + .000
= .035

Yes; since .035 < .05, "at least 3 sevens" would be considered an unusual result.

21. Use probabilities to identify unusual results.

P(10 or more girls) = P(x≥10)
= P(x=10 or x=11 or x=12 or x=13 or x=14)
= P(x=10) + P(x=11) + P(x=12) + P(x=13) + P(x=14)
= .061 + .022 + .006 + .001 + .000
= .090

No; since .090 > .05, "10 or more girls" would not be considered an unusual result for 14 births. The gender selection technique may be effective, but this is not evidence to conclude that it is. While results like those observed might be expected if the technique is effective, they could still reasonably (i.e., with probability of 5% or more) occur if the technique is not effective.

23. The 16 equally like outcomes in the sample space are given at the right. If x represents the number of girls, counting the numbers of favorable outcomes indicates

outcome	x	outcome	x
BBBB	0	GBBB	1
BBBG	1	GBBG	2
BBGB	1	GBGB	2
BGBB	1	GGBB	2
BGGB	2	GGGB	3
BGBG	2	GGBG	3
BBGG	2	GBGG	3
BGGG	3	GGGG	4

P(x = 0) = 1/16 = .0625
P(x = 1) = 4/16 = .2500
P(x = 2) = 6/16 = .3750
P(x = 3) = 4/16 = .2500
P(x = 4) = 1/16 = .0625

x	P(x)	x·P(x)	x²	x²P(x)	
0	.0625	0	0	0	μ = Σx·P(x)
1	.2500	.2500	1	.2500	= 2.0000, rounded to 2.0
2	.3750	.7500	4	1.5000	σ² = Σx²·P(x) − μ²
3	.2500	.7500	9	2.2500	= 5.0000 − (2.0000)²
4	.0625	.2500	16	1.0000	= 1.0000
	1.0000	2.0000		5.0000	σ = 1.0000, rounded to 1.0

4-3 Binomial Experiments

NOTE: The four requirements for a binomial experiment are
#1 There are a fixed number of trials.
#2 The trials are independent.
#3 Each trial has two possible named outcomes.
#4 The probabilities remain constant for each trial.

1. Yes; all four requirements are met [assuming each multiple choice question has the same number of choices, only one of which is correct].

3. No; requirement #3 is not met. There are more than 2 possible outcomes.

5. No; requirement #3 is not met. There are more than 2 possible outcomes.

7. No; requirement #3 is not met. There are more than 2 possible outcomes.

9. let W = guessing the wrong answer
    C = guessing the correct answer
   P(W) = 4/5, for each question
   P(C) = 1/5, for each question
   a. P(WWC) = P(W_1 and W_2 and C_3)
             = P(W_1)·P(W_2)·P(C_3)
             = (4/5)·(4/5)·(1/5)
             = 16/125 = .128
   b. There are 3 possible arrangements: WWC, WCW, CWW
      Following the pattern in part (a)
      P(WWC) = P(W_1 and W_2 and C_3) = P(W_1)·P(W_2)·P(C_3) = (4/5)·(4/5)·(1/5) = 16/125
      P(WCW) = P(W_1 and C_2 and W_3) = P(W_1)·P(C_2)·P(W_3) = (4/5)·(1/5)·(4/5) = 16/125
      P(CWW) = P(C_1 and W_2 and W_3) = P(C_1)·P(W_2)·P(W_3) = (1/5)·(4/5)·(4/5) = 16/125
   c. P(exactly one correct answer) = P(WWC or WCW or CWW)
                                    = P(WWC) + P(WCW) + P(CWW)
                                    = 16/125 + 16/125 + 16/125
                                    = 48/125 = .384

11. from table A-1 in the .01 column and the 7-1 row, .066

13. from table A-1 in the .60 column and the 8-3 row, .124

15. from table A-1 in the .01 column and the 11-4 row, .000$^+$

NOTE: To use the binomial formula, one must identify 3 quantities: n,x,p. Table A-1, for example, requires only these 3 values to supply a probability. Since what the text calls "q" always equals 1-p, it can be so designated without introducing unnecessary notation [just as no special notation is utilized for the quantity n-x, even though it appears twice in the binomial formula]. This has the additional advantage of ensuring that the probabilities p and 1-p sum to 1.00 and protecting against an error in the separate calculation and/or identification of "q." In addition, reversing the order of (n-x)! and x! in the denominator of the $_nC_x$ coefficient term seems appropriate. That agrees with the $n!/(n_1!n_2!)$ logic of the "permutation rule when some objects are alike" for n_1 objects of one type and n_2 objects of another type, and that places the "x" and "n-x" in the same order in both the denominator of the coefficient term and the exponents. Such a natural ordering also leads to fewer errors. Accordingly, this manual expresses the binomial formula as

$$P(x) = [n!/x!(n-x)!]·p^x·(1-p)^{n-x}$$

17. $P(x) = [n!/x!(n-x)!]·p^x·(1-p)^{n-x}$
    $P(x=3) = [5!/3!2!]·(.25)^3·(.75)^2$
            $= [10]·(.25)^3·(.75)^2 = [10]·(.0156)·(.5625) = .0879$

IMPORTANT NOTE: The intermediate values of 10, .0156 and .5625 are given to help those with incorrect answers to identify the portion of the problem in which the mistake was made. This practice will be followed in all problems (i.e., not just binomial problems) throughout the manual. In practice, all calculations can be done in one step on the calculator. You may choose to (or be asked to) write down such intermediate values for your own (or the instructor's) benefit, but <u>never round off in the middle of a problem</u>. <u>Do not write the values down on paper and then re-enter them in the calculator -- use the memory to let the calculator remember with complete accuracy any intermediate values that will be used in subsequent calculations.</u> In addition, always make certain that the quantity [n!/x!(n-x)!] is a whole number and that the final answer is between 0 and 1.

19. $P(x) = [n!/x!(n-x)!] \cdot p^x \cdot (1-p)^{n-x}$
$P(x=4) = [10!/4!6!] \cdot (1/3)^4 \cdot (2/3)^6$
$\qquad = [210] \cdot (1/81) \cdot (64/729)$
$\qquad = .228$

21. $P(x \geq 4) = P(x=4 \text{ or } x=5)$
$\qquad = P(x=4) + P(x=5)$
$\qquad = .1956 + .0459$
$\qquad = .2415$

23. $P(x>1) = 1 - P(x \leq 1)$
$\qquad = 1 - P(x=0 \text{ or } x=1)$
$\qquad = 1 - [P(x=0) + P(x=1)]$
$\qquad = 1 - [.0206 + .1209]$
$\qquad = 1 - [.1415]$
$\qquad = .8585$

NOTE: This also could have been directly as $P(x>1) = P(x=2 \text{ or } x=3 \text{ or } x=4 \text{ or } x=5)$. In general the manual will choose the most efficient technique for solving problems that may be approached in more than one way.

25. let x = the number of taxpayers that are audited
binomial problem: n = 5 and p = .01, use Table A-1
   a. $P(x=3) = .000^+$
   b. $P(x \geq 3) = P(x=3) + P(x=4) + P(x=5)$
   $\qquad = .000^+ + .000^+ + .000^+$
   $\qquad = .000^+$
   c. Since $.000^+ < .05$, "three or more are audited" is an unusual result and we conclude that factors other than chance are at work. It appears that either customers of the Hemingway Company are being targeted by the IRS for audits or that customers with the types of returns that are often audited are selecting the Hemingway Company to do their returns.
   NOTE: A classic scenario of this type is the hospital that has higher than average patient mortality. This can be due either to the care being below average or to the accepting of a higher than average proportion of more serious cases.

27. let x = the number of wrong answers
binomial problem: n = 10 and p = .15, use the binomial formula
   a. $P(x) = [n!/x!(n-x)!] \cdot p^x \cdot (1-p)^{n-x}$
   $P(x=1) = [10!/1!9!] \cdot (.15)^1 \cdot (.85)^9$
   $\qquad = [10] \cdot (.1500) \cdot (.2316)$
   $\qquad = .347$
   b. $P(x=0) = [10!/0!10!] \cdot (.15)^0 \cdot (.85)^{10}$
   $\qquad = [1] \cdot (1) \cdot (.1969)$
   $\qquad = .197$
   $P(x \leq 1) = P(x=0) + P(x=1)$
   $\qquad = .197 + .347$
   $\qquad = .544$
   c. No; since $.544 > .05$, "at most one wrong number" is not an unusual occurrence when the error rate is 15%.

29. let x = the number of men that are color blind
binomial problem: n = 6 and p = .09, use the binomial formula
$P(x) = [n!/x!(n-x)!] \cdot p^x \cdot (1-p)^{n-x}$
$P(x=2) = [6!/2!4!] \cdot (.09)^2 \cdot (.91)^4$
$= [15] \cdot (.0081) \cdot (.6857)$
$= .0833$

31. let x = the number of special program students who graduated
binomial problem: n = 10 and p = .94, use the binomial formula
$P(x) = [n!/x!(n-x)!] \cdot p^x \cdot (1-p)^{n-x}$
a. $P(x \geq 9) = P(x=9) + P(x=10)$
$= [10!/9!1!] \cdot (.94)^9 \cdot (.06)^1 + [10!/10!0!] \cdot (.94)^{10} \cdot (.06)^0$
$= [10] \cdot (.5730) \cdot (.06) + [1] \cdot (.5386) \cdot (1)$
$= .344 + .539$
$= .883$
b. $P(x=8) = [10!/8!2!] \cdot (.94)^8 \cdot (.06)^2$
$= [36] \cdot (.6096) \cdot (.0036) = .079$
$P(x \leq 7) = 1 - P(x > 7)$
$= 1 - [P(x=8) + P(x=9) + P(x=10)]$
$= 1 - [.079 + .344 + .539]$
$= 1 - [.962]$
$= .038$

Yes; since $P(x \leq 7) = .038 < .05$, getting only 7 that graduated would be an unusual result. Recall that "x successes among n trials is unusually low if P(x or fewer) < .05."

33. let x = the number of correct answers
binomial problem: n = 10 and p = .20, use Table A-1
$P(60\% \text{ or higher}) = P(x \geq 6)$
$= P(x=6) + P(x=7) + P(x=8) + P(x=9) + P(x=10)$
$= .006 + .001 + 0^+ + 0^+ + 0^+$
$= .007$

No, the probability is not high enough to make random guessing a reasonable substitute for studying.

35. let x = the number of defective bulbs
binomial problem: n = 24 and p = .04, use the binomial formula
$P(x) = [n!/x!(n-x)!] \cdot p^x \cdot (1-p)^{n-x}$
$P(\text{accept shipment}) = P(x \leq 1)$
$= P(x=0) + P(x=1)$
$= [24!/0!24!] \cdot (.04)^0 \cdot (.96)^{24} + [24!/1!23!] \cdot (.04)^1 \cdot (.96)^{23}$
$= [1] \cdot (1) \cdot (.3754) + [24] \cdot (.04) \cdot (.3911)$
$= .375 + .375$
$= .750$

37. let x = the number of components tested to find 1st defect
geometric problem: p = .2, use the geometric formula
$P(x) = p \cdot (1-p)^{x-1}$
$P(x=7) = (.2) \cdot (.8)^6$
$= (.2) \cdot (.2621)$
$= .0524$

39. Extending the pattern to cover 6 types of outcomes, where $\Sigma x = n$ and $\Sigma p = 1$,
$$P(x_1,x_2,x_3,x_4,x_5,x_6) = [n!/(x_1!x_2!x_3!x_4!x_5!x_6!)]\cdot p_1^{x_1}\cdot p_2^{x_2}\cdot p_3^{x_3}\cdot p_4^{x_4}\cdot p_5^{x_5}\cdot p_6^{x_6}$$
$n = 20$
$p_1 = p_2 = p_3 = p_4 = p_5 = p_6 = 1/6$
$x_1 = 5,\ x_2 = 4,\ x_3 = 3,\ x_4 = 2,\ x_5 = 3,\ x_6 = 3$
Use the multinomial formula.
$$\begin{aligned}
P(x_1,x_2,x_3,x_4,x_5,x_6) &= [n!/(x_1!x_2!x_3!x_4!x_5!x_6!)]\cdot p_1^{x_1}\cdot p_2^{x_2}\cdot p_3^{x_3}\cdot p_4^{x_4}\cdot p_5^{x_5}\cdot p_6^{x_6} \\
&= [20!/(5!4!3!2!3!3!)]\cdot(1/6)^5\cdot(1/6)^4\cdot(1/6)^3\cdot(1/6)^2\cdot(1/6)^3\cdot(1/6)^3 \\
&= [20!/(5!4!3!2!3!3!)]\cdot(1/6)^{20} \\
&= [1.955\cdot10^{12}]\cdot(2.735\cdot10^{-16}) \\
&= .000535
\end{aligned}$$

4.4 Mean, Variance and Standard Deviation for the Binomial Distribution

1. $\mu = n\cdot p = (100)\cdot(.25) = 25.0$
   $\sigma^2 = n\cdot p\cdot(1-p) = (100)\cdot(.25)\cdot(.75) = 18.75$
   $\sigma = 4.3$
   minimum usual value $= \mu - 2\sigma = 25.0 - 2\cdot(4.3) = 16.4$
   maximum usual value $= \mu + 2\sigma = 25.0 + 2\cdot(4.3) = 33.6$

3. $\mu = n\cdot p = (237)\cdot(2/3) = 158.0$
   $\sigma^2 = n\cdot p\cdot(1-p) = (237)\cdot(2/3)\cdot(1/3) = 52.67$
   $\sigma = 7.3$
   minimum usual value $= \mu - 2\sigma = 158.0 - 2\cdot(7.3) = 143.4$
   maximum usual value $= \mu + 2\sigma = 158.0 + 2\cdot(7.3) = 172.6$

5. let x = the number of correct answers
   binomial problem: $n = 20$, $p = .5$
   a. $\mu = n\cdot p = (20)\cdot(.5) = 10.0$
      $\sigma^2 = n\cdot p\cdot(1-p) = (20)\cdot(.5)\cdot(.5) = 5.00$
      $\sigma = 2.2$
   b. maximum usual value $= \mu + 2\sigma = 10.0 + 2\cdot(2.2) = 14.4$
      No; since 12 is less than or equal to the maximum usual value, it would not be unusual for a student to pass by getting at least 12 correct answers.

7. let x = the number of wins
   binomial problem: $n = 100$, $p = 1/38$
   a. $\mu = n\cdot p = (100)\cdot(1/38) = 2.6$
      $\sigma^2 = n\cdot p\cdot(1-p) = (100)\cdot(1/38)\cdot(37/38) = 2.56$
      $\sigma = 1.6$
   b. minimum usual value $= \mu - 2\sigma = 2.6 - 2\cdot(1.6) = -0.6$ [use 0, by practical constraints]
      No; since 0 is greater than or equal to the minimum usual value, it would not be unusual not to win once in the 100 trials.

9. let x = the number of high school graduates
   binomial problem: $n = 1250$, $p = .82$
   a. $\mu = n\cdot p = (1250)\cdot(.82) = 1025.0$
      $\sigma^2 = n\cdot p\cdot(1-p) = (1250)\cdot(.82)\cdot(.18) = 184.50$
      $\sigma = 13.6$
   b. maximum usual value $= \mu + 2\sigma = 1025.0 + 2\cdot(13.6) = 1052.2$
      Yes; since 1107 is greater than the maximum usual value, it would be unusual to get that many high school graduates in a random sample of Americans. But while the 1250

people surveyed were a random sample of Dutchess County residents, they are not a random sample of all Americans -- and the 82% figure was based on all Americans. Since Americans tend to live and associate with others of their same socio-economic class, most communities are probably well above or well below the 82% figure -- with only relatively few communities near the nationwide average.

11. let x = the number testing positive for the HIV virus
    binomial problem: n = 150, p = .10
    a. μ = n·p = (150)·(.10) = 15.0
       σ^2 = n·p·(1-p) = (150)·(.10)·(.90) = 13.50
       σ = 3.7
    b. minimum usual value = μ - 2σ = 15.0 - 2·(3.7) = 7.6
       No; since 12 is greater than or equal to the minimum usual value, it would not be unusual to get that many persons with the HIV virus if the program had no effect and the infection rate were still 10%. The result does not give enough evidence to conclude that the program is effective.

13. let x = the number of customers filing complaints
    binomial problem: n = 850, p = .032
    a. μ = n·p = (850)·(.032) = 27.2
       σ^2 = n·p·(1-p) = (850)·(.032)·(.968) = 26.33
       σ = 5.1
    b. minimum usual value = μ - 2σ = 27.2 - 2·(5.1) = 17.0
       Yes; since 7 is less than the minimum usual value, it would be unusual to get that many complaints if the program had no effect and the rate were still 3.2%. The result does give enough evidence to conclude that the program is effective in lowering the rate of complaints.

15. let x = the number of commercials that are one minute long
    binomial problem: n = 50, p = .06
    a. μ = n·p = (50)·(.06) = 3.0
       σ^2 = n·p·(1-p) = (50)·(.06)·(.94) = 2.82
       σ = 1.7
    b. maximum usual value = μ + 2σ = 3.0 + 2·(1.7) = 6.4
       Yes; since 16 is greater than the maximum usual value, it would be unusual to get that many one-minute commercials if the 6% still applied. The recent result suggests that one-minute commercials are now more than 6% of the total.

17. let x = the number of edible pizzas
    binomial problem: n is unknown, p = .8, we want P(x≥5) ≥ .99
    for n=5, P(x≥5) = P(x=5)
                    = .328
    for n=6, P(x≥5) = P(x=5) + P(x=6)
                    = .393 + .262 = .655
    for n=7, P(x≥5) = P(x=5) + P(x=6) + P(x=7)
                    = .275 + .367 + .210 = .852
    for n=8, P(x≥5) = P(x=5) + P(x=6) + P(x=7) + P(x=8)
                    = .147 + .294 + .336 + .168 = .945
    for n=9, P(x≥5) = P(x=5) + P(x=6) + P(x=7) + P(x=8) + P(x=9)
                    = .066 + .176 + .302 + .302 + .134 = .980
    for n=10, P(x≥5)= P(x=5) + P(x=6) + P(x=7) + P(x=8) + P(x=9) + P(x=10)
                    = .026 + .088 + .201 + .302 + .268 + .107 = .992
    The minimum number of pizzas necessary to be at least 99% sure that there will be 5 edible pizzas available is n = 10.
    NOTE: Because mathematical formulation of this problem involves complicated algebra and does not promote better understanding of the concepts involved, we suggest a trial-

and-error approach using Table A-1. The procedure given above may not be the most efficient, but it is easy to follow and promotes better understanding of the concepts involved.

Review Exercises

1. a. A random variable is a characteristic that assumes a single value (usually a numerical value), determined by chance, for each outcome of an experiment.
   b. A probability distribution is a statement of the possible values a random variable can assume and the probability associated with each of those values. To be valid it must be true for each value x in the distribution that $0 \leq P(x) \leq 1$ and that $\Sigma P(x) = 1$.
   c Yes, the given table is a valid probability distribution because it meets the definition and conditions in part (b) above.

   NOTE: The following table summarizes the calculations for parts (d) and (e).

x	P(x)	x·P(x)	x²	x²P(x)
0	.0004	0	0	0
1	.0094	.0094	1	.0094
2	.0870	.1740	4	.3480
3	.3562	1.0686	9	3.2058
4	.5470	2.1880	16	8.7520
	1.0000	3.4400		12.3152

   d. $\mu = \Sigma x \cdot P(x) = 3.4400 = 3.4$ (rounded)
   e. $\sigma^2 = \Sigma x^2 \cdot P(x) - \mu^2 = 12.3152 - (3.4400)^2 = .4816$
      $\sigma = .7$
   f. Yes; since $P(x=0) = .0004 < .05$.

2. let x = the number of TV's tuned to *ER*
   binomial problem: n = 20 and p = .34, use the binomial formula
   a. $E(x) = \mu = n \cdot p = (20) \cdot (.34) = 6.8$
   b. $\mu = n \cdot p = (20) \cdot (.34) = 6.8$
   c. $\sigma^2 = n \cdot p \cdot (1-p) = (20) \cdot (.34) \cdot (.66) = 4.488$
      $\sigma = 2.1$
   d. $P(x) = [n!/x!(n-x)!] \cdot p^x \cdot (1-p)^{n-x}$
      $P(x=5) = [20!/5!15!] \cdot (.34)^5 \cdot (.66)^{15}$
      $= [15504] \cdot (.0045435) \cdot (.0019641)$
      $= .138$
   e. maximum usual value $= \mu + 2\sigma = 6.8 + 2 \cdot (2.1) = 11$
      Yes; since 12 is greater than the maximum usual value it would be unusual to find that 12 sets are tuned to *ER*.

3. let x = the number of companies that test for drug abuse
   binomial problem: n = 10 and p = .80, use Table A-1
   a. $P(x=5) = .026$
   b. $P(x \geq 5) = P(x=5) + P(x=6) + P(x=7) + P(x=8) + P(x=9) + P(x=10)$
      $= .026 + .088 + .201 + .302 + .268 + .107$
      $= .992$
   c. $\mu = n \cdot p = (10) \cdot (.80) = 8.0$
      $\sigma^2 = n \cdot p \cdot (1-p) = (10) \cdot (.80) \cdot (.20) = 1.60$
      $\sigma = 1.3$
   d. No; since $P(x=6) = .088$, which is not unusual. A particular result is unusual if P(that result or a more extreme result) < .05.

4. let x = the number of workers fired for inability to get along with others
    binomial problem: n = 5 and p = .17, use the binomial formula
    $P(x) = [n!/x!(n-x)!] \cdot p^x \cdot (1-p)^{n-x}$

    a. $P(x=4) = [5!/4!1!] \cdot (.17)^4 \cdot (.83)^1 = (5) \cdot (.000835) \cdot (.83) = .00347$
        $P(x=5) = [5!/5!0!] \cdot (.17)^5 \cdot (.83)^0 = (1) \cdot (.000142) \cdot (1) = .000142$
        $P(x \geq 4) = P(x=4) + P(x=5)$
        $= .00347 + .00014$
        $= .00361$

    b. Yes; x=4 would be an unusual result for a company with the standard rate of 17%. A
        particular result is unusual if P(that result or a more extreme result) < .05, and
        $P(x \geq 4) = .00361 < .05$.

Cumulative Review Exercises

1. a. The table below at the left was used to calculate the mean and standard deviation.

x	f	f·x	f·x²		x	r.f.
0	55	0	0		0	.786
1	2	2	2		1	.029
2	1	2	4		2	.014
3	1	3	9		3	.014
4	0	0	0		4	.000
5	3	15	75		5	.043
6	0	0	0		6	.000
7	2	14	98		7	.029
8	4	32	256		8	.057
9	2	18	162		9	.029
	70	86	606			1.000

$\bar{x} = [\Sigma(f \cdot x)]/[\Sigma f] = 86/70 = 1.2$
$s^2 = \{[\Sigma f] \cdot [\Sigma(f \cdot x^2)] - [\Sigma(f \cdot x)]^2\}/\{[(\Sigma f)] \cdot [(\Sigma f) - 1]\}$
$= [(70) \cdot (606) - (86)^2]/[(70) \cdot (69)]$
$= 37091/4830$
$= 7.251$
$s = 2.7$

    b. The table is given above at the right and was constructed using r.f. = $f/(\Sigma f)$ = f/70.

    c. The first two columns of the table below constitute the requested probability
        distribution.
        The other columns were added to calculate the mean and the standard deviation.

x	P(x)	x·P(x)	x²	x²P(x)
0	.1	0	0	0
1	.1	.1	1	.1
2	.1	.2	4	.4
3	.1	.3	9	.9
4	.1	.4	16	1.6
5	.1	.5	25	2.5
6	.1	.6	36	3.6
7	.1	.7	49	4.9
8	.1	.8	64	6.4
9	.1	.9	81	8.1
	1	4.5		28.5

$\mu = \Sigma x \cdot P(x)$
$= 4.5$
$\sigma^2 = \Sigma x^2 \cdot P(x) - \mu^2$
$= 28.5 - (4.5)^2$
$= 8.25$
$\sigma = 2.8723$, rounded to 2.9

    d. The presence of so many 0's make it clear that the last digits do not represent a random
        sample. Factors other than accurate measurement (which, like social security numbers
        and other such data, would represent random final digits) determined the recorded
        distances.

2. a. The summary statistics are n = 20, Σx = 193, Σx^2 = 2027

$\bar{x}$ = $(\Sigma x)/n$ = 193/20 = 9.65

s^2 = $[n \cdot (\Sigma x^2) - (\Sigma x)^2]/[n \cdot (n-1)]$

= $[20 \cdot (2027) - (193)^2]/[20 \cdot (19)]$

= 3291/380

= 8.66

s = 2.9

b. Of the 20 sample results, 8 were 12's. Use the relative frequency to estimate the probability.

estimated P(x=12) = 8/20 = .400

correct P(x=12) = 1/36 = .0278

These results do not agree. It appears that the dice are not balanced.

c. let x = the number of 12's

binomial problem: n = 20 and p = 1/36, use the binomial formula

$P(x)$ = $[n!/x!(n-x)!] \cdot p^x \cdot (1-p)^{n-x}$

$P(x=0)$ = $[20!/0!20!] \cdot (1/36)^0 \cdot (35/36)^{20}$

= $(1) \cdot (1) \cdot (.569)$

= .569

$P(x \geq 1)$ = 1 - P(x<1)

= 1 - P(x=0)

= 1 - .569

= .431

d. The results obtained in court would be difficult to dispute. Here are some possible ideas and where they lead.

** The probability of at least one 12 is reasonable -- .431, from part (c). Maybe the probability of eight 12's isn't as low as one would guess.

binomial problem: n = 20 and p = 1/36, use the binomial formula

$P(x)$ = $[n!/x!(n-x)!] \cdot p^x \cdot (1-p)^{n-x}$

$P(x=8)$ = $[20!/8!12!] \cdot (1/36)^8 \cdot (35/36)^{12}$

= $(125.970) \cdot (3.54 \cdot 10^{-13}) \cdot (.713)$

= .0000000318

Oops! That's one calculation that the defense definitely shouldn't show the court. Even using P(x≥8), since the rarity of an event is P(that result or a more extreme value), would not be helpful -- as the probabilities for more than 8 12's are even smaller and will not accumulate to any significant value.

** Perhaps the criterion for what constitutes an unusual event would be helpful. For a balanced pair of dice, the court has agreed that μ = 7.0 and σ = 2.4.

Unusual values are those outside $\mu \pm 2 \cdot \sigma$

7.0 ± 2·(2.4)

7.0 ± 4.8

2.2 to 11.8

The sample average in part (a) was $\bar{x}$ = 9.65. Since this is well within these limits, does that mean the result from the accused person's dice should not be considered unusual?

Unfortunately, not. The above limits are for <u>one</u> x value, not for the <u>mean</u> of 20 x values. Rolling the dice <u>once</u> and getting a 9 or 10 would not be considered unusual. The statement says nothing about rolling a pair of dice 20 times and getting a <u>mean</u> of 9 or 10. [The Central Limit Theorem in the following chapter does address this concern, but the results would be very damaging to the accused.]

** The only hope appears to be to complain in non-statistical language that the sample is too small with a diversionary speech such as the following: "We all know that these dice could be rolled millions and millions of times. Using a sample of size 20 to reach a conclusion about all the millions and millions of possible results within these dice is just as foolish as allowing the opinions of a haphazard sample of size 20 to speak for the millions and millions of law-abiding citizens in this great country."

Chapter 5

Normal Probability Distributions

5-2 The Standard Normal Distribution

1. The height of the rectangle is .2. Probability
corresponds to area, and the area of a rectangle
is (width)·(height).
$P(x < 4)$ = (width)·(height)
= (4-0)·(.2)
= (4)·(.2)
= .8

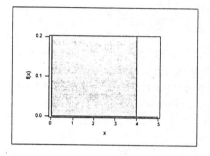

3. The height of the rectangle is .2. Probability
corresponds to area, and the area of a rectangle
is (width)·(height).
$P(3 < x < 4)$ = (width)·(height)
= (4-3)·(.2)
= (1)·(.2)
= .2

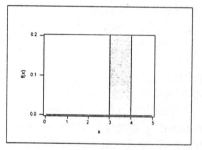

5. The height of the rectangle is 1/6. Probability
corresponds to area, and the area of a rectangle
is (width)·(height).
$P(x > 10)$ = (width)·(height)
= (12-10)·(1/6)
= (2)·(1/6)
= .333

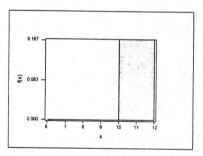

7. The height of the rectangle is 1/6. Probability
corresponds to area, and the area of a rectangle
is (width)·(height).
$P(7 < x < 10)$ = (width)·(height)
= (10-7)·(1/6)
= (3)·(1/6)
= .5

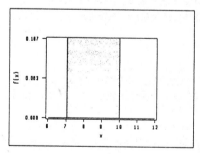

NOTE: The sketch is the key to exercises 9-32. It tells whether to add two Table A-2 probabilities, to subtract two Table A-2 probabilities, to subtract a Table A-2 probability from .5000, to add a Table A-2 probability to .5000, etc. It also often provides a check against gross errors by indicating at a glance whether the final probability is less than or greater than .5000. Remember that the symmetry of the normal curve implies two important facts:

    * There is always .5000 above and below the middle (i.e., at z = 0).

    * $P(-a < z < 0) = P(0 < z < a)$ for all values of "a."

For the remainder of chapter 5, THE ACCOMPANYING SKETCHES ARE NOT TO SCALE and are intended only as aides to help the reader understand how to use the tabled values to answers the questions.

9. $P(0 < z < 1.50)$
      $= .4332$

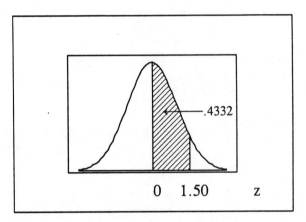

11. $P(-1.96 < z < 0)$
      $= P(0 < z < 1.96)$
      $= .4750$

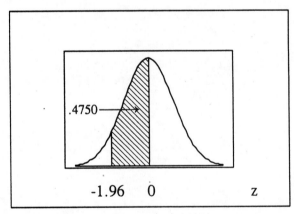

13. $P(z < -1.79)$
      $= P(z < 0) - P(-1.79 < z < 0)$
      $= .5000 - .4633$
      $= .0367$

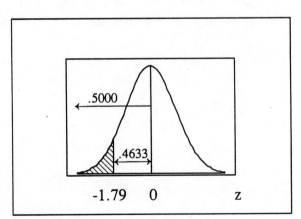

15. P(z > 2.05)
     = P(z > 0) - P(0 < z < 2.05)
     = .5000 - .4798
     = .0202

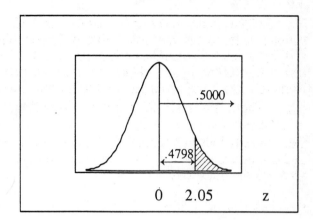

17. P(.50 < z < 1.50)
    = P(0 < z < 1.50) - P(0 < z < .50)
    = .4332 - .1915
    = .2417

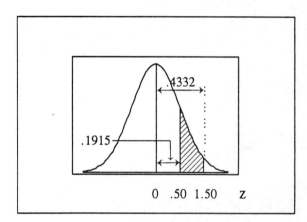

19. P(-2.00 < z < -1.00)
    = P(-2.00 < z < 0) - P(-1.00 < z < 0)
    = .4772 - .3413
    = .1359

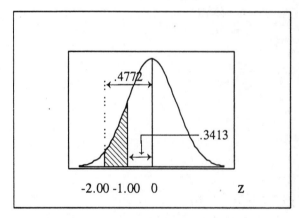

21. P(z < 1.62)
    = P(z < 0) + P(0 < z < 1.62)
    = .5000 + .4474
    = .9474

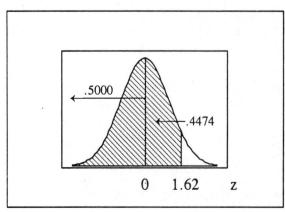

23. P(z > -.27)
    $$= P(-.27 < z < 0) + P(z > 0)$$
    $$= .1064 + .5000$$
    $$= .6064$$

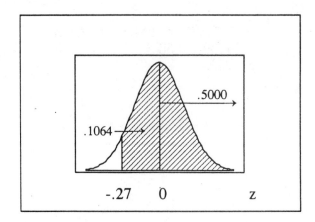

25. P(-1.08 < z < .33)
    $$= P(-1.08 < z < 0) + P(0 < z < .33)$$
    $$= .3599 + .1293$$
    $$= .4892$$

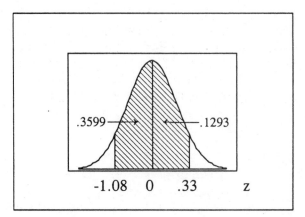

27. P(z > 0)
    $$= .5000$$

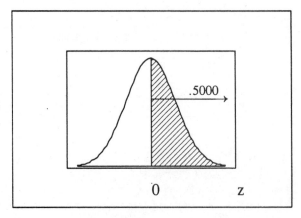

29. P(-1.96 < z < 1.96)
    $$= P(-1.96 < z < 0) + P(0 < z < 1.96)$$
    $$= .4750 + .4750$$
    $$= .9500$$

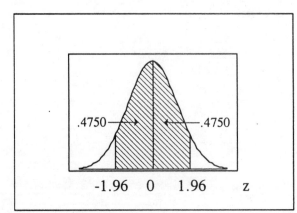

31. $P(z > -2.575)$
    $= P(-2.575 < z < 0) + P(z > 0)$
    $= .4950 + .5000$
    $= .9950$

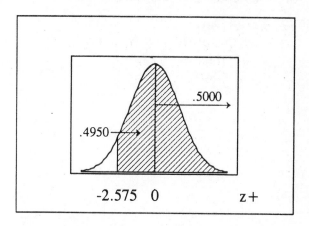

NOTE: The sketch is the key to exercises 33-40. It tells what probability is between 0 and the z score of interest (i.e. the area A to look up when reading Table A-2 "backwards." It also provides a check against gross errors by indicating at a glance whether a z score is above or below 0. Remember that the symmetry of the normal curve implies two important facts·
  * There is always .5000 above and
    below the middle (i.e., at $z = 0$).
  * $P(-a < z < 0) = P(0 < z < a)$ for
    all values of "a."

33. For P_{90}, $A = .4000$.
    The closest entry is $A = .3997$,
        for which $z = 1.28$
    [positive, since it is to the right of the center,
        where $z = 0$].

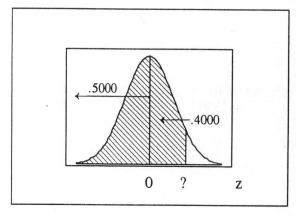

35. For Q_1, $A = .2500$.
    The closest entry is $A = .2486$,
        for which $z = -.67$
    [negative, since it is to the left of the center,
        where $z = 0$].

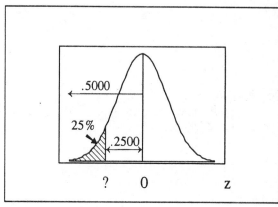

37. For the lowest 5%, A = .4500.
    The closest entry is A = .4500,
       for which z = -1.645
    [negative, since it is to the left of the center,
       where z = 0].

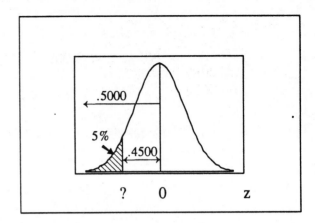

39. For the lowest 3%, A = .4700.
    The closest entry is A = .4699,
       for which z = -1.88
    [negative, since it is to the left of the center,
       where z = 0].

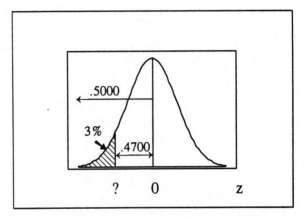

41. Rewrite each of the given statements in terms of z, recalling that z is the number of standard
    deviations a score is from the mean.
    a. $P(-1.00 < z < 1.00)$
       $= P(-1.00 < z < 0)$
             $+ P(0 < z < 1.00)$
       $= .3413 + .3413$
       $= .6826$

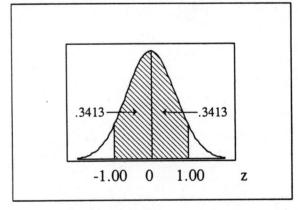

    b. $P(-1.96 < z < 1.96)$
       $= P(-1.96 < z < 0)$
             $+ P(0 < z < 1.96)$
       $= .4750 + .4750$
       $= .9500$

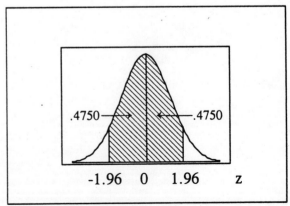

c. P(-3.00 < z < 3.00)
    = P(-3.00 < z < 0)
        + P(0 < z < 3.00)
    = .4987 + .4987
    = .9974

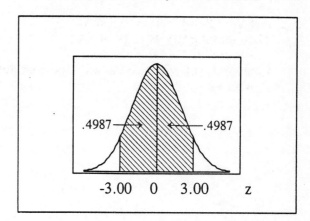

d. P(-1.00 < z < 2.00)
    = P(-1.00 < z < 0)
        + P(0 < z < 2.00)
    = .3413 + .4772
    = .8185

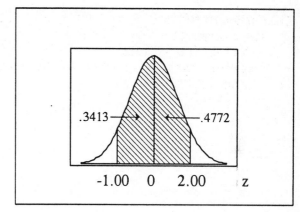

e. P(z < -2.00 or z > 2.00)
    = P(z < -2.00)   + P(z > 2.00)
    = [P(z < 0) - P(-2.00 < z < 0)] + [P(z > 0) - P(0 < z < 2.00)]
    = [.5000 - .4772]   + [.5000 - .4772]
    = .0228   + .0228
    = .0456

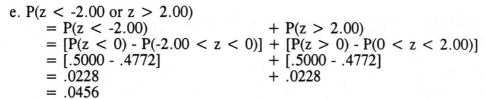

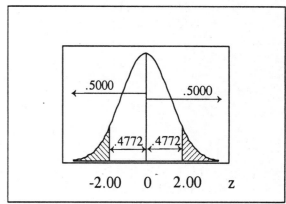

5-3 Nonstandard Normal Distributions: Finding Probabilities

NOTE: In each nonstandard normal distribution, x scores are converted to z scores using the formula $z = (x-\mu)/\sigma$ and rounded to two decimal places. As in the previous section, drawing and labeling the sketch is the key to successful completion of the exercises.

1. $\mu = 143$
   $\sigma = 29$
   $P(143 < x < 172)$
   $= P(0 < z < 1.00)$
   $= .3413$

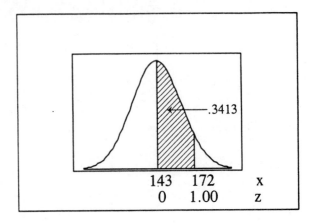

3. $\mu = 143$
   $\sigma = 29$
   $P(x > 150)$
   $= P(z > .24)$
   $= .5000 - .0948$
   $= .4052$

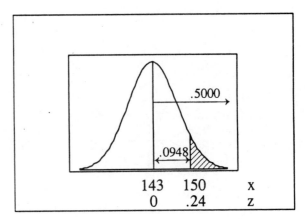

5. $\mu = 63.6$
   $\sigma = 2.5$
   $P(65.5 < x < 68.0)$
   $= P(.76 < z < 1.76)$
   $= .4608 - .2764$
   $= .1844$

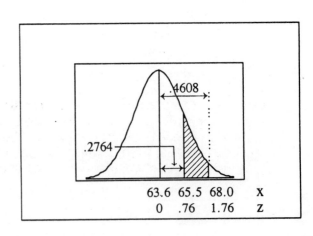

7. $\mu = 63.6$
   $\sigma = 2.5$
   $P(64.5 < x < 72.0)$
   $= P(.36 < z < 3.36)$
   $= .4999 - .1406$
   $= .3593$

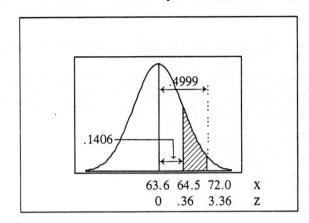

9. $\mu = 998$
   $\sigma = 202$
   $P(x < 900)$
   $= P(z < -.49)$
   $= .5000 - .1879$
   $= .3121$

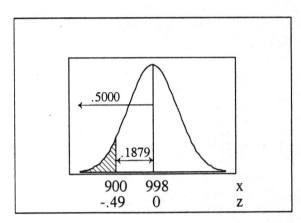

11. $\mu = 3.586$
    $\sigma = .074$
    $P(x < 3.5)$
    $= P(z < -1.16)$
    $= .5000 - .3770$
    $= .1230$

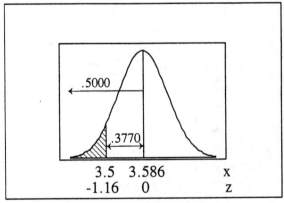

No; it is not true that many consumers are being cheated. Since .1230 > .05, it is not unusual for a packet to contain less than the indicated amount. As consumers do not purchase individual packets, however, they are not really being cheated. When a package of several packets is purchased, the number that are overweight more than compensates for those that are underweight, so that the overall purchase contains more than the indicated amount.

13. $\mu = 268$
    $\sigma = 15$
    $P(x \geq 308)$
    $= P(z \geq 2.67)$
    $= .5000 - .4962$
    $= .0038$

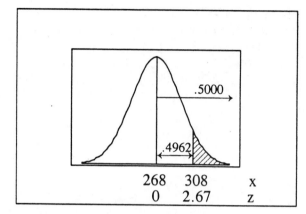

Since .0038 < .05, such a pregnancy is rare -- but certainly not impossible. Out of every one million pregnancies, for example, we expect 3,800 to last 308 days or longer.

15. $\mu = 12.19$
    $\sigma = .11$
    $P(x < 12)$
    $= P(z < -1.73)$
    $= .5000 - .4852$
    $= .0418$

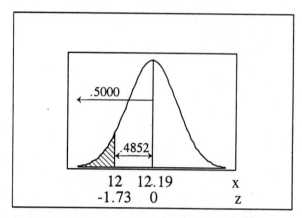

No; it is not true that many consumers are being cheated. Since .0418 < .05, getting a can with less than 12 ounces is unusual. And while a "defect" rate of 4% may seem high for a reputable product, the facts that $\mu > 12$ and that most people buy several cans over a period of time means that the consumer still comes out ahead in the long run.

17. $\mu = 69.0$
    $\sigma = 2.8$
    $P(64 < x < 78)$
    $= P(-1.79 < z < 3.21)$
    $= .4633 + .4999$
    $= .9632$

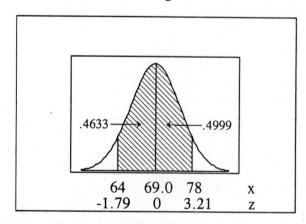

No; it is not true that height prevents too many men from joining the Marines.

19. $\mu = 3570$
$\quad \sigma = 500$
$\quad P(x < 2200)$
$\quad\quad = P(z < -2.74)$
$\quad\quad = .5000 - .4969$
$\quad\quad = .0031$

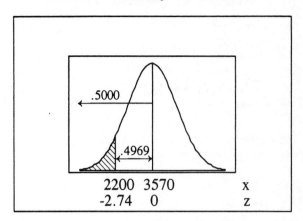

The percentage of babies in the "at-risk" category is .31%. If a hospital experiences 900 births a year, we expect about 3 of them to be "at-risk."

21. $\mu = 178.1$
$\quad \sigma = 40.7$
$\quad P(x > 300)$
$\quad\quad = P(z > 3.00)$
$\quad\quad = .5000 - .4987$
$\quad\quad = .0013$

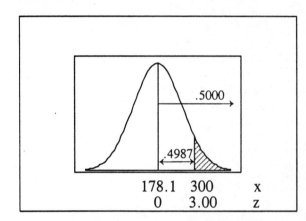

This probability may not cause an individual male aged 18 to 24 to be concerned. It may be, however, that the medical community should be concerned. In a metropolitan area of 100,000 such people, for example, we expect about 130 to be "at-risk."

23. $\mu = 100$
$\quad \sigma = 15$
$\quad$ a. $P(x > 131.5)$
$\quad\quad = P(z > 2.10)$
$\quad\quad = .5000 - .4821$
$\quad\quad = .0179$

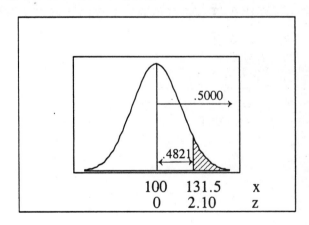

$\quad$ b. $(.0179) \cdot (75,000) = 1,342.5$

25. No matter what the original problem, z scores have no units. Consider, for example, the z score for a weight of 150 from the population of women used in this section.

$$z = \frac{150 \text{ lb} - 143 \text{ lb}}{29 \text{ lb}} = \frac{7 \text{ lb}}{29 \text{ lb}} = .24, \text{ the "lb" units having "cancelled out"}$$

5-4 Nonstandard Normal Distributions: Finding Scores

NOTE: The relationship between standardized and raw scores is given by $z = (x-\mu)/\sigma$. That formula may be solved, as needed, for any of the quantities on the right to yield
$$x = \mu + z\sigma \qquad \sigma = (x-\mu)/z \qquad \mu = x - z\sigma$$

As in the previous section, drawing and labeling the sketch is the key to successful completion of the exercises. Remember that the z score for the mean is zero, and that z scores to the left of the mean must be negative.

1. $\mu = 10$
   $\sigma = 2$
   for P_{90}, A = .4000 [.3997] and z = 1.28
   $x = \mu + z\sigma$
   $\quad = 10 + (1.28)(2)$
   $\quad = 10 + 2.6$
   $\quad = 12.6$

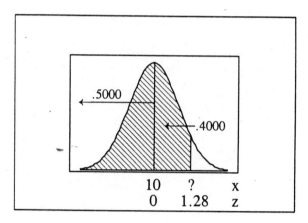

3. $\mu = 10$
   $\sigma = 2$
   for D_2, A = .3000 [.2995] and z = -.84
   $x = \mu + z\sigma$
   $\quad = 10 + (-.84)(2)$
   $\quad = 10 - 1.7$
   $\quad = 8.3$

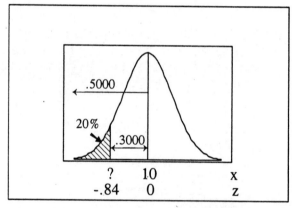

5. $\mu = 143$
   $\sigma = 29$
   for Q_3, A = .2500 [.2486] and z = .67
   $x = \mu + z\sigma$
   $\quad = 143 + (.67)(29)$
   $\quad = 143 + 19$
   $\quad = 162$ lbs

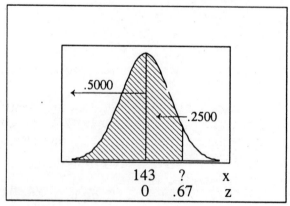

7. $\mu = 143$
   $\sigma = 29$
   for P_{15}, A = .3500 [.3508] and z = -1.04
   $x = \mu + z\sigma$
   $= 143 + (-1.04)(29)$
   $= 143 - 30$
   $= 113$ lbs

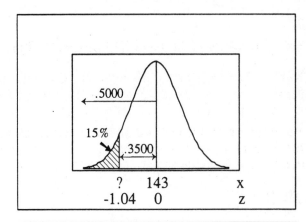

9. $\mu = 63.6$
   $\sigma = 2.5$
   for the tallest 5%, A = .4500 and z = 1.645
   $x = \mu + z\sigma$
   $= 63.6 + (1.645)(2.5)$
   $= 63.6 + 4.1$
   $= 67.7$ inches

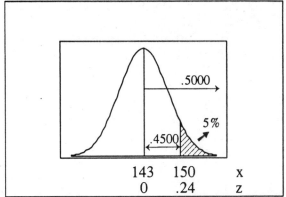

"A height that is in the top 5% of all women" does not identify a specific value for women to know whether they are eligible to apply. If the population continues to grow taller, as it has over the past decades, the membership requirement will not be constant and will require monitoring the population in some way.

11. $\mu = 1017$
    $\sigma = 207$
    for the top 42%,
       A = .0800 [.0793] and z = .20
    $x = \mu + z\sigma$
    $= 1017 + (.20)(207)$
    $= 1017 + 41$
    $= 1058$

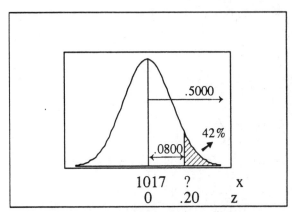

"A score in the top 42%" does not identify a specific value that students can aim for and/or determine whether they are eligible to apply. It may generate undesirable negative interaction if students perceive that they are competing against each other and not against some standard.

13. $\mu = 8.2$
    $\sigma = 1.1$
    a. P(x < 5)
       = P(z < -2.91)
       = .5000 - .4982
       = .0018

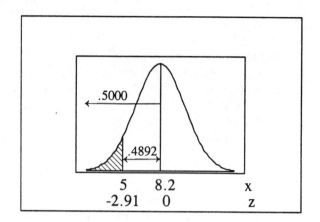

    b. for the poorest 1%,
       A = .4900 [.4901] and z = -2.33
       x = μ + zσ
         = 8.2 + (-2.33)(1.1)
         = 8.2 - 2.6
         = 5.6 years

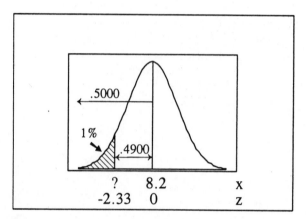

In practice, the warranty time would be rounded (down) to a "nice" amount -- probably 5.5 years or (equivalently) 66 months.

15. $\mu = 9.4$
    $\sigma = 4.2$
    a. P(x < 15)
       = P(z < 1.33)
       = .5000 + .4082
       = .9082

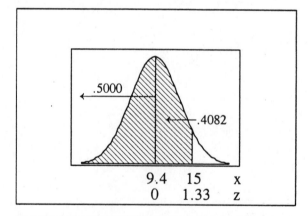

    b. for P_{99}, A = .4900 [.4901] and z = 2.33
       x = μ + zσ
         = 9.4 + (2.33)(4.2)
         = 9.4 + 9.8
         = 19.2 lbs

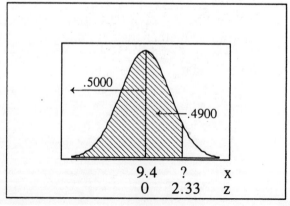

17. $\mu = 268$
    $\sigma = 15$
    for the lowest 4%,
       $A = .4600\ [.4599]$ and $z = -1.75$.
    $x = \mu + z\sigma$
       $= 268 + (-1.75)(15)$
       $= 268 - 26$
       $= 242$ days

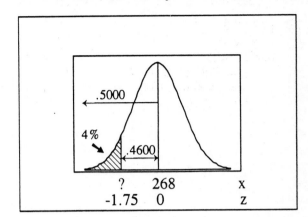

19. $\mu = 178.1$
    $\sigma = 40.7$
    a. $P(x > 250)$
       $= P(z > 2.01)$
       $= .5000 - .4778$
       $= .0222$

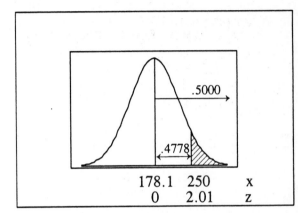

b. for the top 3%,
       $A = .4700\ [.4699]$ and $z = 1.88$
    $x = \mu + z\sigma$
       $= 178.1 + (1.88)(40.7)$
       $= 178.1 + 76.5$
       $= 254.6$ mg/100 mL

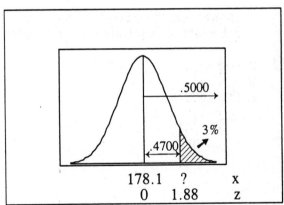

21. $\mu = 143$
    $\sigma = 29$
    a. $\mu \pm 2\sigma = 143 \pm 2 \cdot (29)$
                $= 143 \pm 58$
                $= 85$ to 201

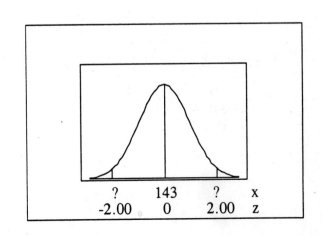

b. P(-2.00 < z < 2.00)
      = .4772 + .4772
      = .9544
      = 95.44%

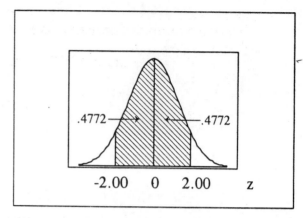

c. for A = .9500/2 = .4750, z = 1.96
    $\mu \pm 1.96\sigma$ = 143 $\pm$ 1.96·(29)
              = 143 $\pm$ 56.84
              = 86.16 to 199.84

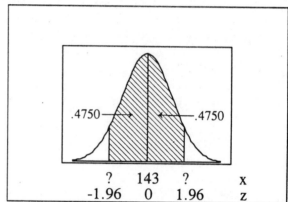

5-5 The Central Limit Theorem

NOTE: When using individual scores (i.e., making a statement about one x score from the original distribution), convert x to z using the mean and standard deviation of the x's and z = (x-μ)/σ.

When using a sample of n scores (i.e., making a statement about), convert $\overline{x}$ to z using the mean and standard deviation of the $\overline{x}$'s and z = ($\overline{x}$-$\mu_{\overline{x}}$)/$\sigma_{\overline{x}}$.

IMPORTANT NOTE: After calculating $\sigma_{\overline{x}}$, <u>STORE IT</u> in the calculator to recall it with total accuracy whenever it is needed in subsequent calculations. <u>DO NOT</u> write it down on paper rounded off (even to several decimal places) and then re-enter it in the calculator whenever it is needed. This avoids both round-off errors and recopying errors.

1. a. normal distribution
        μ = 143
        σ = 29
        P(143 < x < 155)
           = P(0 < z < .41)
           = .1591

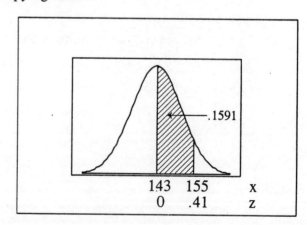

b. normal distribution,
   since the original distribution is so
   $\mu_{\bar{x}} = \mu = 143$
   $\sigma_{\bar{x}} = \sigma/\sqrt{n} = 29/\sqrt{36} = 4.833$
   $P(143 < \bar{x} < 155)$
   $= P(0 < z < 2.48)$
   $= .4934$

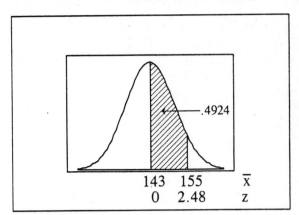

NOTE: Remember that the figures are for illustrative purposes, as an aide to solving the problem, and are not to scale. In scaled drawings, the shaded area would be very narrow in part 1(a) and very wide in part 1(b).

3. a. normal distribution
   $\mu = 143$
   $\sigma = 29$
   $P(x > 160)$
   $= P(z > .59)$
   $= .5000 - .2224$
   $= .2776$

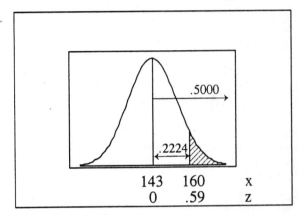

b. normal distribution,
   since the original distribution is so
   $\mu_{\bar{x}} = \mu = 143$
   $\sigma_{\bar{x}} = \sigma/\sqrt{n} = 29/\sqrt{50} = 4.101$
   $P(\bar{x} > 160)$
   $= P(z > 4.15)$
   $= .5000 - .4999$
   $= .0001$

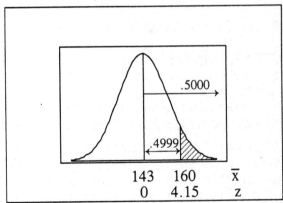

5. a. normal distribution,
   since the original distribution is so
   $\mu_{\bar{x}} = \mu = 143$
   $\sigma_{\bar{x}} = \sigma/\sqrt{n} = 29/\sqrt{16} = 7.250$
   $P(150 < \bar{x} < 155)$
   $= P(.97 < z < 1.66)$
   $= .4515 - .3340$
   $= .1175$

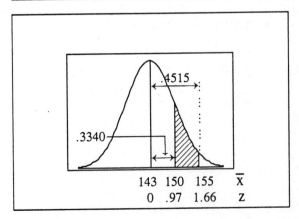

b. When the original distribution is normal, the sampling distribution of the $\bar{x}$'s is normal regardless of the sample size.

7. normal distribution,
   by the Central Limit Theorem
   $\mu_{\bar{x}} = \mu = .500$
   $\sigma_{\bar{x}} = \sigma/\sqrt{n} = .289/\sqrt{50} = .0409$
   $P(.6 < \bar{x} < .7)$
       $= P(2.45 < z < 4.89)$
       $= .4999 - .4929$
       $= .0070$

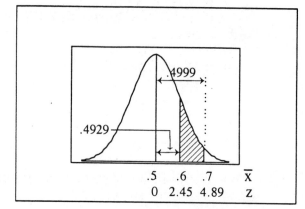

Yes; since .0070 < .05, it would be unusual for a properly-functioning computer random number generator to generate 50 such numbers.

9. a. normal distribution,
   by the Central Limit Theorem
   $\mu_{\bar{x}} = \mu = 8.2$
   $\sigma_{\bar{x}} = \sigma/\sqrt{n} = 1.1/\sqrt{50} = .156$
   $P(\bar{x} \leq 7.8)$
       $= P(z \leq -2.57)$
       $= .5000 - .4949$
       $= .0051$

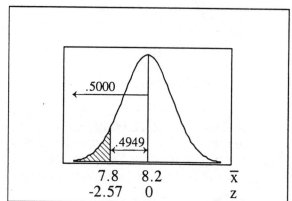

b. Yes; it appears that the TV sets sold by the Portland Electronics store are of less than average quality.

11. a. normal distribution
    $\mu = 32.473$
    $\sigma = 5.601$
    $P(x < 29.000)$
        $= P(z < -.62)$
        $= .5000 - .2324$
        $= .2676$

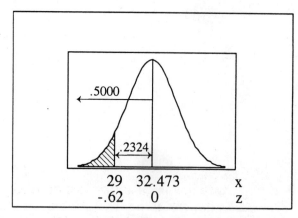

b. NOTE: Completion of the problem assumes that "a decade of ten years" means 10 years chosen at random and used to form an artificial "decade." Since weather may have long range cycles and patterns (so that the rainfall in any one year is not independent of the rainfall in adjacent years), a chronological decade could not be assumed to represent a random sample equivalent to 10 randomly selected years.

normal distribution,
  since the original distribution is so
$\mu_{\bar{x}} = \mu = 32.473$
$\sigma_{\bar{x}} = \sigma/\sqrt{n} = 5.601/\sqrt{10} = 1.771$
$P(\bar{x} < 29.00)$
  $= P(z < -1.96)$
  $= .5000 - .4750$
  $= .0250$

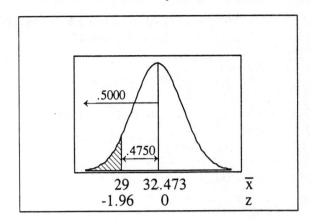

c. When the original distribution is normal, the sampling distribution of the $\bar{x}$'s is normal regardless of the sample size.

13. a. normal distribution,
  by the Central Limit Theorem
$\mu_{\bar{x}} = \mu = 12.00$
$\sigma_{\bar{x}} = \sigma/\sqrt{n} = .09/\sqrt{36} = .015$
$P(\bar{x} \geq 12.29)$
  $= P(z \geq 19.33)$
  $= .5000 - .4999$
  $= .0001$

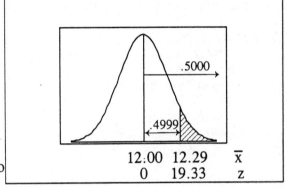

b. It appears that the cans are NOT being filled with a mean of 12.00 oz and a standard deviation of .09 oz. Since the cans appear to contain more than 12.00 oz, and that poses no difficulties for consumers, consumers are not being cheated.

15. a. normal distribution
$\mu = 509$
$\sigma = 112$
$P(x \geq 590)$
  $= P(z \geq .72)$
  $= .5000 - .2642$
  $= .2358$

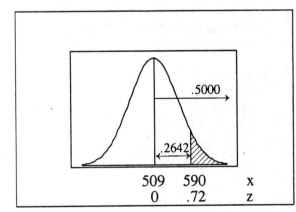

b. normal distribution,
  since the original distribution is so
$\mu_{\bar{x}} = \mu = 509$
$\sigma_{\bar{x}} = \sigma/\sqrt{n} = 112/\sqrt{16} = 28$
$P(\bar{x} \geq 590)$
  $= P(z \geq 2.89)$
  $= .5000 - .4981$
  $= .0019$

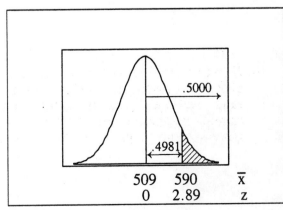

c. When the original distribution is normal, the sampling distribution of the $\bar{x}$'s is normal regardless of the sample size.

d. Yes; since .0019 < .05 by quite a bit, there is strong evidence to support the claim that the course is effective -- assuming, as stated, that the 16 participants were truly a random sample (and not self-selected or otherwise chosen in some biased manner).

17. normal distribution,
     since the original distribution is so
     $\mu_{\bar{x}} = \mu = 27.44$
     $\sigma_{\bar{x}} = \sigma/\sqrt{n} = 12.46/\sqrt{4872} = .179$
     $P(\bar{x} > 27.88)$
          $= P(z > 2.46)$
          $= .5000 - .4931$
          $= .0069$

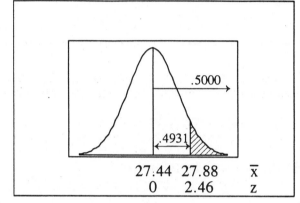

The system is currently acceptable and can expect to be overloaded only about $(.0069)(52) = .36$ weeks a year -- or approximaely once every three years.

19. $\mu = .9147$
     $\sigma = .0369$
     a. This cannot be answered without knowing the shape of the distribution.
        NOTE: If the weights of the M&M's are normally distributed, then Table A-2 can be used and $P(x > .9085) = P(z > -.17) = .0675 + .5000 = .5675$.

b. normal distribution,
     by the Central Limit Theorem
     $\mu_{\bar{x}} = \mu = .9147$
     $\sigma_{\bar{x}} = \sigma/\sqrt{n} = .0369/\sqrt{1498} = .000953$
     $P(\bar{x} > .9085)$
          $= P(z > -6.50)$
          $= .4999 + .5000$
          $= .9999$

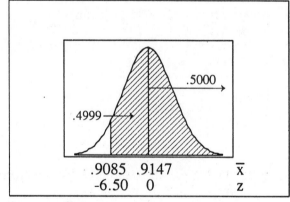

c. Yes; the company is providing at least the amount claimed on the label with almost absolute certainty.

21. normal distribution, since the original distribution is so
     NOTE: Since $16/300 = .0533 > .05$, use the finite population correction factor
     $\mu_{\bar{x}} = \mu = 143$
     $\sigma_{\bar{x}} = [\sigma/\sqrt{n}] \cdot \sqrt{(N-n)/(N-1)}$
          $= [29/\sqrt{16}] \cdot \sqrt{(300-16)/(300-1)}$
          $= [29/\sqrt{16}] \cdot \sqrt{(284)/(299)}$
          $= 7.066$
     $P(\bar{x} > 157.5)$
          $= P(z > 2.05)$
          $= .5000 - .4798$
          $= .0202$

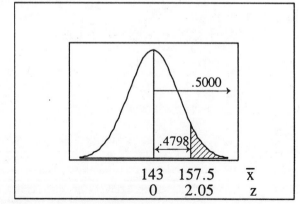

Since $(12) \cdot (365) \cdot (.0202) = 88.48$, the elevator is likely to be overloaded about 88 times a year -- i.e., about once every 4 days. If the weights of the members of the Atlanta Women's Club are distributed as the given population, the club should consider limiting the elevator's capacity to 15.

5-6 Normal Distribution as Approximation to Binomial Distribution

NOTE: As in the previous sections, P(E) represents the probability of an event E; this manual uses $P_c(E)$ to represent the probability of an event E with the continuity correction applied.

1. the area to the right of 57.5
      in symbols, $P(x > 57) = P_c(x > 57.5)$

3. the area to the left of 56.5
      in symbols, $P(x < 57) = P_c(x < 56.5)$

5. the area to the left of 24.5
      in symbols, $P(x \le 24) = P_c(x < 24.5)$

7. the area from 15.5 to 22.5
      in symbols, $P(16 \le x \le 22) = P_c(15.5 < x < 22.5)$

IMPORTANT NOTE: As in the previous sections, store σ in the calculator so that it may be recalled with complete accuracy whenever it is needed in subsequent calculations.

9. binomial: $n = 10$ and $p = .80$
      a. from Table A-1, $P(x = 7) = .201$
      b. normal approximation <u>not</u> appropriate since
            $n(1-p) = 10(.20) = 2 < 5$

11. binomial: $n = 15$ and $p = .60$
      a. from Table A-1,
          $P(x \ge 9) = P(x = 9) + P(x = 10) + \ldots + P(x = 15)$
                    $= .207 + .186 + .127 + .063 + .022 + .005 + 0^+$
                    $= .610$
      b. normal approximation appropriate since
            $np = 15(.60) = 9 \ge 5$
            $n(1-p) = 15(.40) = 6 \ge 5$
            $\mu = np = 15(.60) = 9$
            $\sigma = \sqrt{np(1-p)} = \sqrt{15(.60)(.40)} = 1.897$
            $P(x \ge 9)$
                $= P_c(x > 8.5)$
                $= P(z > -.26)$
                $= .1026 + .5000 = .6026$

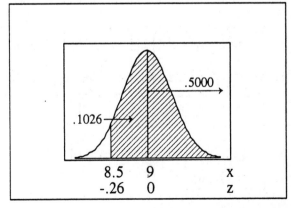

13. let x = the number of girls born
binomial: n = 100 and p = .50
normal approximation appropriate since
    np = 100(.50) = 50 ≥ 5
    n(1-p) = 100(.50) = 50 ≥ 5
μ = np = 100(.50) = 50
$\sigma = \sqrt{np(1-p)} = \sqrt{100(.50)(.50)} = 5.000$
P(x ≥ 52)
    = P_c(x > 51.5)
    = P(z > .30)
    = .5000 - .1179 = .3821

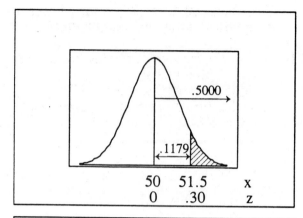

15. let x = the number of correct responses
binomial: n = 50 and p = .50
normal approximation appropriate since
    np = 50(.50) = 25 ≥ 5
    n(1-p) = 50(.50) = 25 ≥ 5
μ = np = 50(.50) = 25
$\sigma = \sqrt{np(1-p)} = \sqrt{50(.50)(.50)} = 3.536$
P(x ≥ 30)
    = P_c(x > 29.5)
    = P(z > 1.27)
    = .5000 - .3980 = .1020

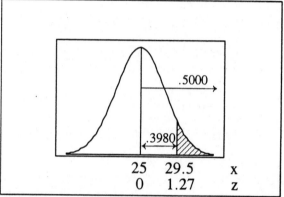

17. let x = the number of wrong numbers
binomial: n = 50 and p = .15
normal approximation appropriate since
    np = 50(.15) = 7.5 ≥ 5
    n(1-p) = 50(.85) = 42.5 ≥ 5
μ = np = 50(.15) = 7.5
$\sigma = \sqrt{np(1-p)} = \sqrt{50(.15)(.85)} = 2.525$
P(x < 3)
    = P_c(x < 2.5)
    = P(z < -1.98)
    = .5000 - .4761 = .0239

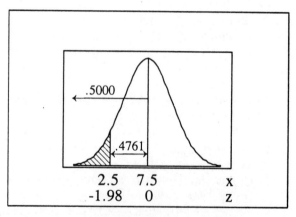

Since .0239 < .05, getting fewer than 3 error would be an unusual event if the error rate were really 15%. Conclude that the error rate for AT&T is less than 15%.

19. let x = the number of graduates
binomial: n = 500 and p = .94
normal approximation appropriate since
    np = 500(.94) = 470 ≥ 5
    n(1-p) = 500(.06) = 30 ≥ 5
μ = np = 500(.94) = 470
$\sigma = \sqrt{np(1-p)} = \sqrt{500(.94)(.06)} = 5.310$
P(x < 456)
    = P_c(x < 455.5)
    = P(z < -2.73)
    = .5000 - .4968 = .0032

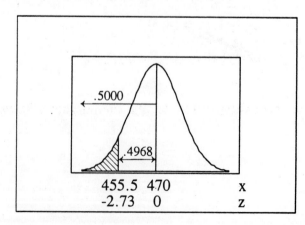

Yes; since .0032 < .05, getting fewer than 456 who graduate would be an unusual event if the graduation rate were really 94%.

21. let x = the number of booked passengers
           that arrive
    binomial: n = 400 and p = .85
    normal approximation appropriate since
        np = 400(.85) = 340 ≥ 5
        n(1-p) = 400(.15) = 60 ≥ 5
    μ = np = 400(.85) = 340
    σ = $\sqrt{np(1-p)}$ = $\sqrt{400(.85)(.15)}$ = 7.141
    P(x > 350)
        = P_c(x > 350.5)
        = P(z > 1.47)
        = .5000 - .4292 = .0708

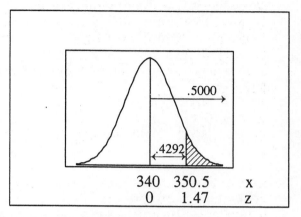

Yes; the airline would probably be willing to allow overbooking to occur 7% of the time (i.e., about 1 in every 14 flights) in order to fly with fewer empty seats.

23. let x = the number of women hired
    binomial: n = 62 and p = .50
    normal approximation appropriate since
        np = 62(.50) = 31 ≥ 5
        n(1-p) = 62(.50) = 31 ≥ 5
    μ = np = 62(.50) = 31
    σ = $\sqrt{np(1-p)}$ = $\sqrt{62(.50)(.50)}$ = 3.937
    P(x ≤ 21)
        = P_c(x < 21.5)
        = P(z < -2.41)
        = .5000 - .4920 = .0080

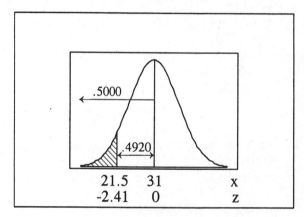

Yes; since the resulting probability is so small, it does support a charge of discrimination based on gender.

25. let x = the number that test positive
    binomial: n = 150 and p = .038
    normal approximation appropriate since
        np = 150(.038) = 5.7 ≥ 5
        n(1-p) = 150(.962) = 144.3 ≥ 5
    μ = np = 150(.038) = 5.7
    σ = $\sqrt{np(1-p)}$ = $\sqrt{150(.038)(.962)}$ = 2.342
    P(x ≥ 10)
        = P_c(x > 9.5)
        = P(z > 1.62)
        = .5000 - .4474 = .0526

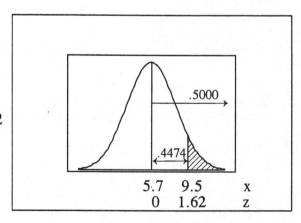

No; since .0526 > .05, the number with positive test results is not unusually high.

27. let x = the number with group O blood
    binomial: n = 400 and p = .45
    normal approximation appropriate since
    $\quad$ np = 400(.45) = 180 ≥ 5
    $\quad$ n(1-p) = 400(.55) = 220 ≥ 5
    μ = np = 400(.45) = 180
    $\sigma = \sqrt{np(1-p)} = \sqrt{400(.45)(.55)} = 9.950$
    P(x ≥ 177)
    $\quad = P_c(x > 176.5)$
    $\quad = P(z > -.35)$
    $\quad = .1368 + .5000 = .6368$

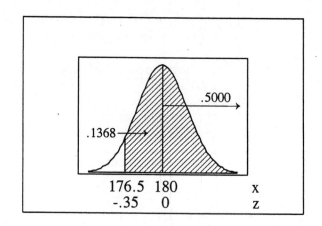

29. let x = the number of times Marc wins $35
    $\qquad$ because 7 occurs
    binomial: n = 200 and p = 1/38
    normal approximation appropriate since
    $\quad$ np = 200(1/38) = 5.26 ≥ 5
    $\quad$ n(1-p) = 200(37/38) = 194.76 ≥ 5
    μ = np = 200(1/38) = 5.26
    $\sigma = \sqrt{np(1-p)} = \sqrt{200(1/38)(37/38)} = 2.264$
    for a profit,
    $\quad$ Marc needs at least 6 wins @ $35 each
    $\quad$ P(x ≥ 6)
    $\qquad = P_c(x > 5.5)$
    $\qquad = P(z > .10)$
    $\qquad = .5000 - .0398 = .4602$

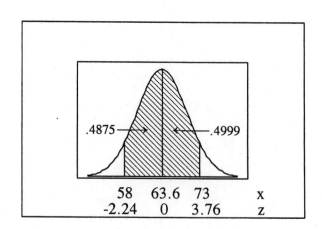

Review Exercises

1. normal: μ = 63.6 and σ = 2.5
   $\quad$ P(58 < x < 73)
   $\qquad = P(-2.24 < z < 3.76)$
   $\qquad = .4875 + .4999$
   $\qquad = .9874$

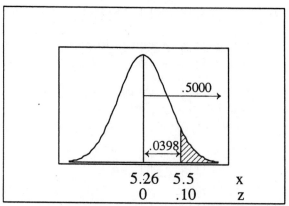

b. For the shortest 1%,

    $A = .4900 [.4901]$ and $z = -2.33$.

$x = \mu + z\sigma$

   $= 63.6 + (-2.33)(2.5)$

   $= 63.6 - 5.8$

   $= 57.8$

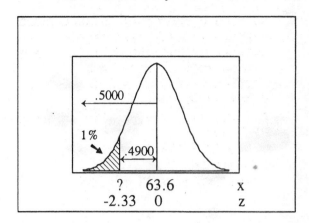

For the tallest 1%,

    $A = .4900 [.4901]$ and $z = 2.33$.

$x = \mu + z\sigma$

   $= 63.6 + (2.33)(2.5)$

   $= 63.6 + 5.8$

   $= 69.4$

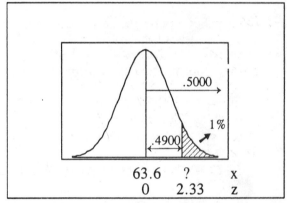

c. normal, since the original distribution is so

  $\mu_{\bar{x}} = \mu = 63.6$

  $\sigma_{\bar{x}} = \sigma/\sqrt{n} = 2.5/\sqrt{9} = .833$

  $P(63.0 < \bar{x}\ 65.0)$

    $= P(-.72 < z < 1.68)$

    $= .2642 + .4535$

    $= .7177$

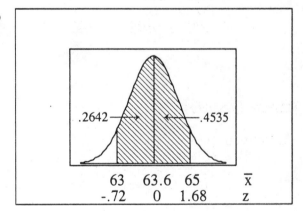

2. normal: $\mu = 69.0$ and $\sigma = 2.8$

  a. $P(x \geq 74)$

    $= P(z > 1.79)$

    $= .5000 - .4633$

    $= .0367$

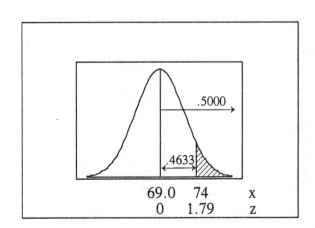

b. P(62.0 < x < 76.0)
    = P(-2.50 < z < 2.50)
    = .4938 + .4938
    = .9876

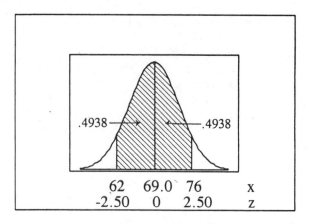

c. For the shortest 2%,
    A = .4800 [4798] and z = -2.05.
    x = μ + zσ
    = 69.0 + (-2.05)(2.8)
    = 69.0 - 5.7
    = 63.3

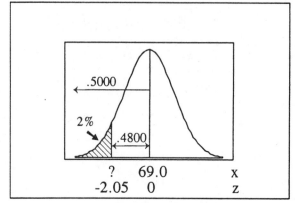

For the tallest 2%,
    A = .4800 [.4798] and z = 2.05.
    x = μ + zσ
    = 69.0 + (2.05)(2.8)
    = 69.0 + 5.7
    = 74.7

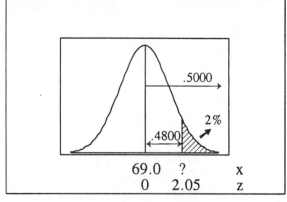

d. normal, since the original distribution is so
    $\mu_{\bar{x}}$ = μ = 69.0
    $\sigma_{\bar{x}}$ = σ/√n = 2.8/√25 = .560
    P($\bar{x}$ > 68.0)
        = P(z > -1.79)
        = .4633 + .5000
        = .9633

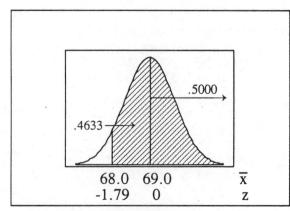

3. normal: $\mu = 132.6$ and $\sigma = 5.4$

   a. $P(x > 140)$

       $= P(z > 1.37)$

       $= .5000 - .4147$

       $= .0853$

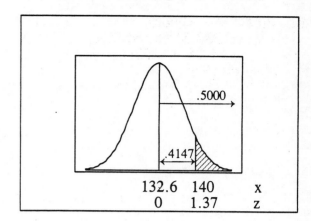

   b. For $D_2 = P_{20}$,

       $A = .3000$ [.2995] and $z = -.84$.

    $x = \mu + z\sigma$

       $= 132.6 + (-.84)(5.4)$

       $= 132.6 - 4.5$

       $= 128.1$

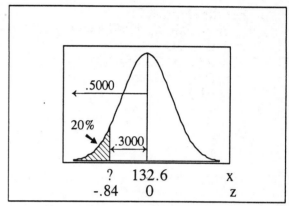

   c. normal, since the original distribution is so

    $\mu_{\bar{x}} = \mu = 132.6$

    $\sigma_{\bar{x}} = \sigma/\sqrt{n} = 5.4/\sqrt{35} = .913$

    $P(133.0 < \bar{x} < 134.0)$

       $= P(.44 < z < 1.53)$

       $= .4370 - .1700$

       $= .2670$

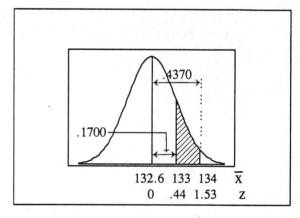

4. binomial: $n = 500$ and $p = .26$

   normal approximation appropriate since

    $np = 500(.26) = 130 \geq 5$

    $n(1-p) = 500(.74) = 370 \geq 5$

   $\mu = np = 500(.26) = 130$

   $\sigma = \sqrt{np(1-p)} = \sqrt{500(.26)(.74)} = 9.808$

   $P(125 \leq x \leq 150)$

       $= P_c(124.5 < x < 150.5)$

       $= P(-.56 < z < 2.09)$

       $= .2123 + .4817$

       $= .6940$

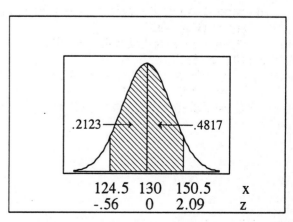

5. normal distribution with $\mu = 0$ and $\sigma = 1$
   [Because this is a standard normal, the variable is already expressed in terms of z scores.]

   a. P(0 < z < 1.00)
       = .3413

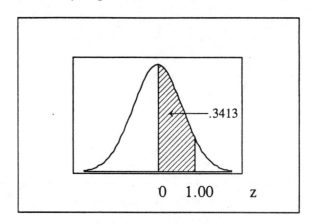

   b. P(z > 2.00)
       = .5000 - .4772
       = .0228

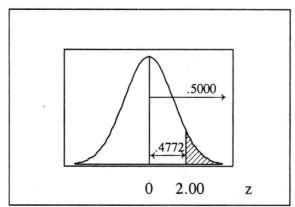

   c. P(z > -1.50)
       = .4332 + .5000
       = .9332

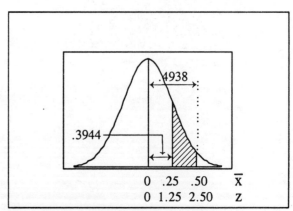

   d. normal, since the original distribution is so
      $\mu_{\bar{x}} = \mu = 0$
      $\sigma_{\bar{x}} = \sigma/\sqrt{n} = 1/\sqrt{25} = .200$
      P(.25 < $\bar{x}$ < .50)
         = P(1.25 < z < 2.50)
         = .4938 - .3944
         = .0994

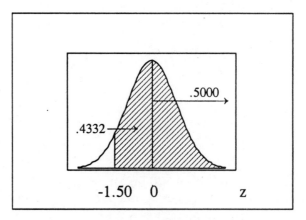

e. For $D_4 = P_{40}$,
   $A = .1000$ [.0987] and $z = -.25$.

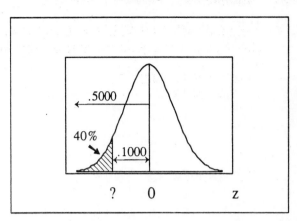

6. uniform distribution. Since the total area must be 1.00 and the width of the rectangle is .4, the height of the rectangle is $1/.4 = 2.5$. Probability corresponds to area, and the area of a rectangle is (width)·(height).

   a. $P(x \geq 12.0)$
      = (width)·(height)
      = (12.2 - 12.0)·(2.5)
      = (.2)·(2.5)
      = .5

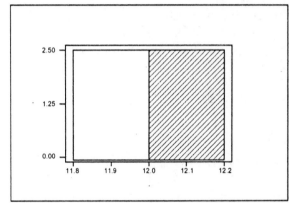

   b. $P(11.9 < x < 12.1)$
      = (width)·(height)
      = (12.1 - 11.9)·(2.5)
      = (.2)·(2.5)
      = .5

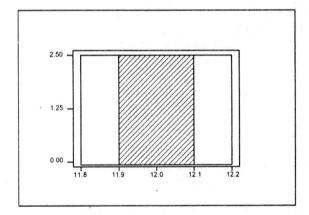

   c. $P(x < 11.9)$
      = (width)·(height)
      = (11.9 - 11.8)·(2.5)
      = (.1)·(2.5)
      = .25

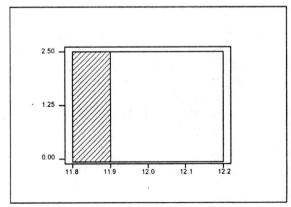

7. normal: $\mu = 9.1$ and $\sigma = 2.1$

a. $P(x > 8.5)$
    $= P(z > -.29)$
    $= .1141 + .5000$
    $= .6141$

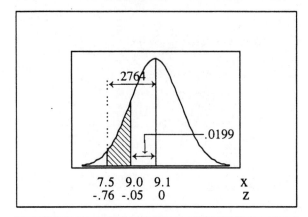

b. $P(7.5 < x < 9.0)$
    $= P(-.76 < z < -.05)$
    $= .2764 - .0199$
    $= .2565$

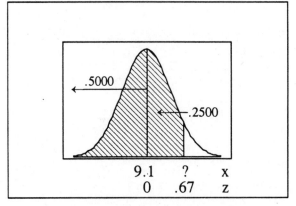

c. For $Q_3 = P_{75}$,
    $A = .2500$ [.2486] and $z = .67$.
    $x = \mu + z\sigma$
    $ = 9.1 + (.67)(2.1)$
    $ = 9.1 + 1.4$
    $ = 10.5$

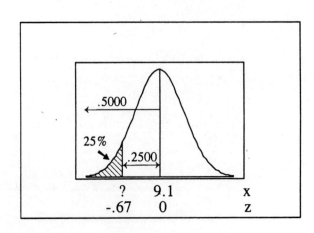

d. For $Q_1 = P_{25}$,
    $A = .2500$ [.2486] and $z = -.67$.
    $x = \mu + z\sigma$
    $ = 9.1 + (-.67)(2.1)$
    $ = 9.1 - 1.4$
    $ = 7.7$

e. normal, since the original distribution is so
$\mu_{\bar{x}} = \mu = 9.1$
$\sigma_{\bar{x}} = \sigma/\sqrt{n} = 2.1/\sqrt{35} = .355$
$P(\bar{x} > 9.0)$
$= P(z > -.28)$
$= .1103 + .5000$
$= .6103$

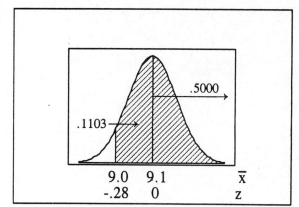

$$\begin{array}{cc} 9.0 & 9.1 \qquad \bar{x} \\ -.28 & 0 \qquad z \end{array}$$

8. binomial: $n = 800$ and $p = .18$
normal approximation appropriate since
$np = 800(.18) = 144 \geq 5$
$n(1-p) = 800(.82) = 656 \geq 5$
$\mu = np = 800(.18) = 144$
$\sigma = \sqrt{np(1-p)} = \sqrt{800(.18)(.82)} = 10.866$
$P(x \geq 150)$
$= P_c(x > 149.5)$
$= P(z > .51)$
$= .5000 - .1950 = .3050$

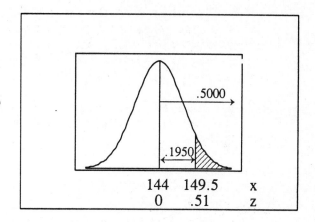

$$\begin{array}{cc} 144 & 149.5 \qquad x \\ 0 & .51 \qquad z \end{array}$$

9. normal: $\mu = 35,600$ and $\sigma = 4275$
a. $P(x > 30,000)$
$= P(z > -1.31)$
$= .4049 + .5000$
$= .9049$

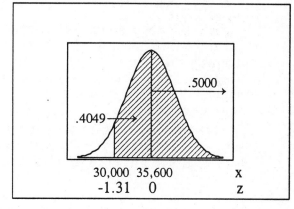

$$\begin{array}{cc} 30,000 & 35,600 \qquad x \\ -1.31 & 0 \qquad z \end{array}$$

b. normal, since the original distribution is so
$\mu_{\bar{x}} = \mu = 35,600$
$\sigma_{\bar{x}} = \sigma/\sqrt{n} = 4275/\sqrt{40} = 675.9$
$P(\bar{x} > 35,000)$
$= P(z > -.89)$
$= .3133 + .5000$
$= .8133$

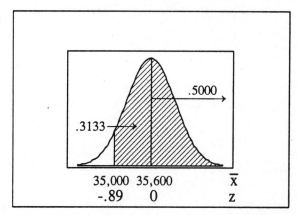

$$\begin{array}{cc} 35,000 & 35,600 \qquad \bar{x} \\ -.89 & 0 \qquad z \end{array}$$

c. For the lowest 3%,
    $A = .4700$ [.4699] and $z = -1.88$.
  $x = \mu + z\sigma$
    $= 35,600 + (-1.88)(4275)$
    $= 35,600 - 8037$
    $= 27,563$

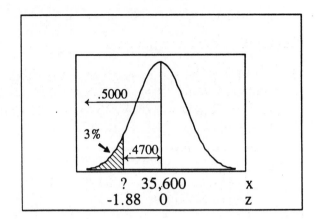

A guarantee of 27,000 or 27,500 miles should be offered.

10. binomial: $n = 36$ and $p = 1/6$
normal approximation appropriate since
  $np = 36(1/6) = 6 \geq 5$
  $n(1-p) = 36(5/6) = 30 \geq 5$
$\mu = np = 36(1/6) = 6$
$\sigma = \sqrt{np(1-p)} = \sqrt{36(1/6)(5/6)} = 2.236$
$P(x \geq 13)$
  $= P_c(x > 12.5)$
  $= P(z > 2.91)$
  $= .5000 - .4982 = .0018$

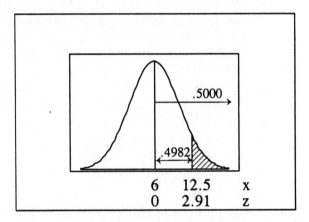

Yes; since $.0018 < .05$, this is strong evidence that he has the ability to predict the outcomes. NOTE: Statistically, the result of 13 or more correct predictions if the assumptions that he was guessing at random is true is very remote -- and so we conclude he was not guessing at random. Cynically, a person may wish to note there were other assumptions (e.g., the die was balanced) -- and perhaps the conclusion should be that one of those is false.

Cumulative Review Exercises

1. A. tally of the scores is given at the right.
   summary values: n = 43, Σx = 2624, Σx^2 = 160,672
   a. $\bar{x}$ = $(\Sigma x)/n$ = (2624)/43 = 61.023 rounded to 61.0 mm
   b. $\tilde{x}$ = 61 mm
   c. M = 60 mm
   d. s^2 = $[n(\Sigma x^2) - (\Sigma x)^2]/[n(n-1)]$
      = $[43(160,672) - (2624)^2]/[43(42)]$
      = 23520/1806
      = 13.023
      s = 3.609 rounded to 3.6 mm
   e. While the tally of the original scores does not appear
      bell-shaped, the frequency distribution given below
      the tally does. Yes; based on the grouped data,
      it does seem that the values are from a population
      with a distribution that is approximately normal.
   f. z = $(x-\bar{x})/s$
      z_{66} = (66 - 61.023)/3.609 = 1.38
   g. r.f. of scores greater than 59 is 31/43 = .7209 = 72%
   h. normal: μ = 61.023 and σ = 3.609
      P(x > 59) = P(z > -.56)
         = .2123 + .5000
         = .7123

52-1	
53-	
54-1	
55-1	
56-111	
57-1	
58-1	
59-1111	
60-111111111	
61-11	
62-11111	
63-11111	
64-11	
65-1	
66-11111	
67-11	

width (mm)	frequency
51 – 53	1
54 – 56	5
57 – 59	6
60 – 62	16
63 – 65	8
66 – 68	7
	43

   i. ratio, since distances between scores are meaningful and
      there is a natural zero
   j. continuous, since length can be any value on a continuum

2. a. Let L = a person is left-handed
      P(L) = .10, for each random selection
      $P(L_1$ and L_2 and $L_3)$ = $P(L_1)\cdot P(L_2)\cdot P(L_3)$ = $(.10)\cdot(.10)\cdot(.10)$ = .001
   b. Let N = a person is not left-handed
      P(N) = .90, for each random selection
      P(at least one left-hander) = 1 - P(no left-handers)
                                  = $1 - P(N_1$ and N_2 and $N_3)$
                                  = $1 - P(N_1)\cdot P(N_2)\cdot P(N_3)$
                                  = 1 - (.90)·(.90)·(.90)
                                  = 1 - .729
                                  = .271
   c. binomial: n = 3 and p = P(left-hander) = .10
      normal approximation <u>not</u> appropriate since
         np = 3(.10) = 0.3 < 5
   d. binomial: n = 50 and p = .10
      μ = np = 50(.10) = 5
   e. binomial: n = 50 and p = .10
      σ = $\sqrt{np(1-p)}$ = $\sqrt{50(.10)(.90)}$ = 2.121
   f. An unusual score is one that is more than 2 standard deviations from the mean.
      Since z = $(x-\mu)/\sigma$, z_8 = (8-5)/2.212 = 1.41.
      Since 8 is 1.41 standard deviations from the mean, it is <u>not</u> an unusual score.

Chapter 6

Estimates and Sample Sizes

6-2 Estimating a Population Mean: Large Samples

1. let L = the lower confidence limit; U = the upper confidence limit
    $\bar{x} = (L + U)/2$
    $= (72.6 + 82.6)/2 = 155.2/2 = 77.6$
    $E = (U - L)/2$
    $= (82.6 - 72.6)/2 = 10.0/2 = 5.0$

3. let L = the lower confidence limit; U = the upper confidence limit
    $\bar{x} = (L + U)/2$
    $= (207 + 212)/2 = 419/2 = 209.5$
    $E = (U - L)/2$
    $= (212 - 207)/2 = 5/2 = 2.5$

5. $\alpha = .01, \alpha/2 = .005$
    $A = .5000 - .005 = .4950$
    $z_{.005} = 2.575$

7. $\alpha = .03, \alpha/2 = .015$
    $A = .5000 - .015 = .4850$
    $z_{.015} = 2.17$

9. n > 30, use z (with s for σ)
    $\alpha = .05, \alpha/2 = .025, z_{.025} = 1.96$
    a. $E = z_{.025} \cdot \sigma/\sqrt{n}$ b. $\bar{x} \pm E$
       $= 1.96 \cdot (12,345)/\sqrt{60}$ $85,678 \pm 3,124$
       $= 3,124$ dollars $82,554 < \mu < 88,802$ (dollars)

11. n > 30, use z (with s for σ)
    $\alpha = .01, \alpha/2 = .005, z_{.005} = 2.33$
    a. $E = z_{.005} \cdot \sigma/\sqrt{n}$ b. $\bar{x} \pm E$
       $= 2.575 \cdot (2.72)/\sqrt{150}$ $7.60 \pm .57$
       $= .57$ seconds $7.03 < \mu < 8.17$ (seconds)

13. We are 95% confident that the interval from 133.35 mm to 135.65 mm contains the true value of μ, the mean maximum skull breadth of all Egyptian males who lived around 1850 B.C.

15. a. n > 30, use z (with s for σ)   b. n > 30, use z (with s for σ)
       $\bar{x} \pm z_{.025} \cdot \sigma/\sqrt{n}$ $\bar{x} \pm z_{.025} \cdot \sigma/\sqrt{n}$
       $205.5 \pm 1.96 \cdot 35.4/\sqrt{40}$ $185.0 \pm 1.96 \cdot 28.4/\sqrt{40}$
       205.5 ± 11.0 185.0 ± 8.8
       $194.5 < \mu < 216.5$ $176.2 < \mu < 193.8$
    c. We are 95% certain that the interval from 194.5 to 216.5 contains the true mean score for oiled lanes, and we are 95% certain that the interval from 176.2 to 193.8 contains the true mean score for dry lanes. Since these intervals do not overlap, we can be fairly certain that the means are truly different and that the lane surface does have an effect on bowling scores -- with the oily surface producing higher scores.
    NOTE: From probability theory, the confidence we have in both statements

simultaneously is $(.95) \cdot (.95) = .9025$ -- about 90% confidence.

17. preliminary values: $n = 36$, $\Sigma x = 5866.98$, $\Sigma x^2 = 1832952.2176$
$\bar{x} = (\Sigma x)/n = (5866.98)/36 = 162.972$
$s^2 = [n(\Sigma x^2) - (\Sigma x)^2]/[n(n-1)] = [36(1832952.2176) - (5866.98)^2]/[36(35)] = 25051.448$
$s = 158.276$
$n > 30$, use $\underline{z}$ (with s for σ)
$\bar{x} \pm z_{.005} \cdot \sigma/\sqrt{n}$
$162.972 \pm 2.575 \cdot 158.276/\sqrt{36}$
162.972 ± 67.927
$95.045 < \mu < 230.898$
NOTE: Here and in similar problems, $\bar{x}$ and s are shown with appropriate accuracy, but the values were stored and recalled <u>with full accuracy</u> for use in subsequent calculations.

19. preliminary values: $n = 40$, $\Sigma x = 38.023$, $\Sigma x^2 = 36.143980$
$\bar{x} = (\Sigma x)/n = (38.023)/40 = .9506$
$s^2 = [n(\Sigma x^2) - (\Sigma x)^2]/[n(n-1)] = [40(36.143980) - (38.023)^2]/[40(39)] = .0000068$
$s = .0026$
$n > 30$, use $\underline{z}$ (with s for σ)
$\bar{x} \pm z_{.05} \cdot \sigma/\sqrt{n}$
$.9506 \pm 1.645 \cdot .0026/\sqrt{40}$
$.9506 \pm .0006$
$.9499 < \mu < .9513$

21. a. preliminary values: $n = 36$, $\Sigma x = 29.4056$, $\Sigma x^2 = 24.02112012$
$\bar{x} = (\Sigma x)/n = (29.4056)/36 = .81682$
$s^2 = [n(\Sigma x^2) - (\Sigma x)^2]/[n(n-1)]$
$\quad = [36(24.02112012) - (29.4056)^2]/[36(35)]$
$\quad = .0000564$
$s = .00751$
$n > 30$, use $\underline{z}$ (with s for σ)
$\bar{x} \pm z_{.025} \cdot \sigma/\sqrt{n}$
$.81682 \pm 1.96 \cdot .00751/\sqrt{36}$
$.81682 \pm .00245$
$.81437 < \mu < .81927$
   b. preliminary values: $n = 36$, $\Sigma x = 28.2526$, $\Sigma x^2 = 22.17315828$
$\bar{x} = (\Sigma x)/n = (28.2526)/36 = .78479$
$s^2 = [n(\Sigma x^2) - (\Sigma x)^2]/[n(n-1)]$
$\quad = [36(22.17315828) - (28.2526)^2]/[36(35)]$
$\quad = .0000193$
$s = .00439$
$n > 30$, use $\underline{z}$ (with s for σ)
$\bar{x} \pm z_{.025} \cdot \sigma/\sqrt{n}$
$.78479 \pm 1.96 \cdot .00439/\sqrt{36}$
$.78479 \pm .00143$
$.78336 < \mu < .78623$
   c. We are 95% certain that the interval from .81437 to .81927 contains the true mean weight for regular Coke, and we are 95% certain that the interval from .78336 to .78623 contains the true mean weight for diet Coke. Since these intervals do not overlap, we can be fairly certain that the means are truly different and that the diet Coke weighs less.
   NOTE: From probability theory, the confidence we have in both statements simultaneously is $(.95) \cdot (.95) = .9025$ -- about 90% confidence.

23. | plastic | paper |
    |---------|-------|

<table>
<tr><td>

plastic
$n = 62$
$\Sigma x = 118.47$
$\Sigma x^2 = 295.6233$
$\bar{x} = (\Sigma x)/n = (118.47)/62 = 1.911$
$s^2 = [n(\Sigma x^2) - (\Sigma x)^2]/[n(n-1)]$
$\quad = [62(295.6233) - (118.47)^2]/[62(61)]$
$\quad = 1.1352$
$s = 1.065$
$n > 30$, use z (with s for σ)
$\bar{x} \pm z_{.04} \cdot \sigma/\sqrt{n}$
$1.911 \pm 1.75 \cdot 1.065/\sqrt{62}$
$1.911 \pm .237$
$1.674 < \mu < 2.148$ (lbs)

</td><td>

paper
$n = 62$
$\Sigma x = 584.54$
$\Sigma x^2 = 6570.8217$
$\bar{x} = (\Sigma x)/n = (584.54)/62 = 9.428$
$s^2 = [n(\Sigma x^2) - (\Sigma x)^2]/[n(n-1)]$
$\quad = [62(6570.8217) - (584.54)^2]/[62(61)]$
$\quad = 17.3728$
$s = 4.168$
$n > 30$, use z (with s for σ)
$\bar{x} \pm z_{.04} \cdot \sigma/\sqrt{n}$
$9.428 \pm 1.75 \cdot 4.168/\sqrt{62}$
$9.428 \pm .926$
$8.502 < \mu < 10.354$ (lbs)

</td></tr>
</table>

By weight, discarded paper is a much larger ecological problem than discarded plastic. Since paper is less dense than plastic, it would also be a larger problem by volume. Since paper is more degradeable and recyclable than plastic, however, it would be difficult to make an unqualified statement.

25. The usual formulas express the two values $\bar{x}$ and s^2 in terms of the two quantities Σx and Σx^2. Solving the usual formulas "in reverse" yields $\Sigma x = n\bar{x}$ and $\Sigma x^2 = (n-1)s^2 + n\bar{x}^2$. In this problem, use the given values of $\bar{x}$ and s^2 to determine Σx and Σx^2, subtract out 98.6 and 98.6^2, and then add in 986 and 986^2.

    original $\Sigma x = n\bar{x}$
    $\qquad = 106 \cdot (98.20) = 10409.2$
    original $\Sigma x^2 = (n-1)s^2 + n\bar{x}^2$
    $\qquad = 105 \cdot (.62)^2 + 106 \cdot (98.20)^2 = 1022223.802$
    new $\Sigma x = 10409.2 - 98.6 + 986$
    $\qquad = 11296.6$
    new $\Sigma x^2 = 1022223.802 - 98.6^2 + 986^2$
    $\qquad = 1984697.84$

    a. $\bar{x} = (\Sigma x)/n = (11296.6)/106 = 106.57$
    $s^2 = [n(\Sigma x^2) - (\Sigma x)^2]/[n(n-1)]$
    $\quad = [106(1984697.84) - (11296.6)^2]/[106(105)]$
    $\quad = 7436.19$
    $s = 86.233$

    $n > 30$, use z (with s for σ)
    $\bar{x} \pm z_{.025} \cdot \sigma/\sqrt{n}$
    $106.57 \pm 1.96 \cdot 86.233/\sqrt{106}$
    106.57 ± 16.42
    $90.16 < \mu < 122.99$

    b. In this problem, the outlier changed the center of the confidence interval from 98.20 to 106.57 and the width of the confidence interval from 98.32-98.08 = .24 to 122.99-90.16 = 32.83. Both changes were significant, and they demonstrate that confidence intervals can be very sensitive to outliers.

    c. Based on part (b), values determined to be outliers should probably not be used in further statistical analyses -- including the construction of confidence intervals.

6-3 Estimating a Population Mean: Small Samples

IMPORTANT NOTE: This manual uses the following conventions.
(1) The designation "df" stands for "degrees of freedom."
(2) Since the t value depends on both the degrees of freedom and the probability lying beyond it, double subscripts are used to identify points on t distributions. The t distribution with 15 degrees of freedom and .025 beyond it, for example, is designated $t_{15,.025} = 2.132$.
(3) When df$\geq$30 the difference between the t and z distributions is negligible and Table A-3 uses one final row of z values to cover all such cases. Consequently, the z scores for certain "popular" α and $\alpha/2$ values may be found by reading Table A-3 "frontwards" instead of reading Table A-2 "backwards." This is not only easier but also more accurate, since Table A-3 includes one more decimal place. The manual uses this technique from this point on.

1. population approximately normal, n $\leq$ 30, σ unknown: use t
   n = 20; df = 19
   α = .05; $\alpha/2$ = .025
   $t_{19,.025} = 2.093$

3. population approximately normal, n $\leq$ 30, σ unknown: use t
   n = 25; df = 14
   α = .01; $\alpha/2$ = .005
   $t_{14,.005} = 2.977$

5. population skewed, n $\leq$ 30: neither the t nor the z applies

7. population approximately normal, n $\leq$ 30, σ unknown: use t
   n = 9; df = 8
   α = .02; $\alpha/2$ = .01
   $t_{8,.01} = 2.896$

9. population approximately normal, n $\leq$ 30, σ unknown: use t
   a. $E = t_{14,.025} \cdot s/\sqrt{n}$
   $= 2.145 \cdot 108/\sqrt{15}$
   $= 60$
   b. $\bar{x} \pm E$
   496 ± 60
   $436 < \mu < 556$

11. $99 < \mu < 109$ [rounded to agree with the accuracy of $\bar{x}$]
    We are 95% certain that the interval from 99 to 109 contains the true mean IQ score for professional athletes.

13. n $\leq$ 30 and σ unknown, use t
    $\bar{x} \pm t_{9,.025} \cdot s/\sqrt{n}$
    $124 \pm 2.262 \cdot 18/\sqrt{10}$
    124 ± 13
    $111 < \mu < 137$ (beats per minute)
    Since the corresponding interval for manual snow shoveling was $164 < \mu < 186$, it appears that the maximum heart rates are lower during automated snow shoveling.

15. n $\leq$ 30 and σ unknown, use t
    $\bar{x} \pm t_{11,.025} \cdot s/\sqrt{n}$
    $26,227 \pm 2.201 \cdot (15,873)/\sqrt{12}$
    $26,227 \pm 10,085$
    $16,142 < \mu < 36,312$ (dollars)
    We are 95% certain that the interval from \$16,142 to \$36,312 contains the true mean repair cost for repairing Dodge Vipers under the specified conditions.

17. preliminary values: n = 7, Σx = .85, Σx^2 = .1123

$\bar{x} = (\Sigma x)/n = (.85)/7 = .121$

$s^2 = [n(\Sigma x^2) - (\Sigma x)^2]/[n(n-1)]$

$\quad = [7(.1123) - (.85)^2]/[7(6)]$

$\quad = .00151$

$s = .039$

n ≤ 30 and σ unknown, use t

$\bar{x} \pm t_{6,.01} \cdot s/\sqrt{n}$

.121 ± 3.143·(.039)/√7

.121 ± .046

.075 < μ < .168

19. n > 30, use z (with s for σ)

$\bar{x} \pm z_{.05} \cdot \sigma/\sqrt{n}$

8.2 ± 1.645·(1.1)/√75

8.2 ± .2

8.0 < μ < 8.4 (years)

No; these sets were made at least 8 years ago, and the technology has changed since then.

21. a. preliminary values: n = 26, Σx = 23.849, Σx^2 = 21.904543

$\bar{x} = (\Sigma x)/n = (23.849)/26 = .9173$

$s^2 = [n(\Sigma x^2) - (\Sigma x)^2]/[n(n-1)]$

$\quad = [26(21.904543) - (23.849)^2]/[26(25)]$

$\quad = .0011436$

$s = .0338$

n ≤ 30 and σ unknown, use t

$\bar{x} \pm t_{25,.05} \cdot s/\sqrt{n}$

.9173 ± 1.708·.0338/√26

.9173 ± .0113

.9059 < μ < .9286 (grams)

b. preliminary values: n = 33, Σx = 30.123, Σx^2 = 27.546793

$\bar{x} = (\Sigma x)/n = (30.123)/33 = .9128$

$s^2 = [n(\Sigma x^2) - (\Sigma x)^2]/[n(n-1)]$

$\quad = [33(27.546793) - (30.123)^2]/[33(32)]$

$\quad = .0015616$

$s = .0395$

n > 30, use z (with s for σ)

$\bar{x} \pm z_{.05} \cdot \sigma/\sqrt{n}$

.9128 ± 1.645·.0395/√33

.9128 ± .0113

.9015 < μ < .9241 (grams)

c. The larger sample size in part (b) allowed the use of the z distribution instead of the t. While both intervals have essentially the same width, the brown M&M's in part (b) may be slightly lighter -- but not significantly so, as the intervals overlap considerably.

23. a. preliminary values: n = 25, Σx = 1972.45, Σx^2 = 160239.3228

$\bar{x} = (\Sigma x)/n = (1972.45)/25 = 78.898$

$s^2 = [n(\Sigma x^2) - (\Sigma x)^2]/[n(n-1)]$

$\quad = [25(160239.3228) - (1972.45)^2]/[25(24)]$

$\quad = 192.37$

$s = 13.87$

$n \leq 30$ and σ unknown, use t
$\bar{x} \pm t_{24,.01} \cdot s/\sqrt{n}$
$78.898 \pm 2.492 \cdot 13.87/\sqrt{25}$
78.898 ± 6.913
$71.985 < \mu < 85.811$ (dollars)

b. preliminary values: $n = 15$, $\Sigma x = 632.25$, $\Sigma x^2 = 30139.2424$
$\bar{x} = (\Sigma x)/n = (632.25)/15 = 42.150$
$s^2 = [n(\Sigma x^2) - (\Sigma x)^2]/[n(n-1)] = [15(30139.2424) - (632.25)^2]/[15(14)] = 249.28$
$s = 15.79$

$n \leq 30$ and σ unknown, use t
$\bar{x} \pm t_{14,.01} \cdot s/\sqrt{n}$
$42.150 \pm 2.625 \cdot 15.79/\sqrt{15}$
42.150 ± 10.701
$31.119 < \mu < 52.851$ (dollars)

c. The paperbacks cost significantly less. Since there was more variability and a smaller sample size for the paperbacks, the confidence interval in part (b) is considerably wider than the one in part (a).

25. For any α, the z value is smaller than the corresponding t value (although the difference decreases as n increases). This creates a smaller E and a narrower confidence interval than one is entitled to -- i.e., it does not take into consideration the extra uncertainty created by using the sample s instead of the true population σ.

27. preliminary values: $n = 5$, $\Sigma x = 44.2$, $\Sigma x^2 = 397.52$
$\bar{x} = (\Sigma x)/n = (44.2)/5 = 8.84$
$s^2 = [n(\Sigma x^2) - (\Sigma x)^2]/[n(n-1)] = [5(397.52) - (44.2)^2]/[5(4)] = 1.698$
$s = 1.30$

$n \leq 30$ and σ unknown, use t
$E = t_{4,\alpha/2} \cdot s/\sqrt{n}$
$1.24 = t_{4,\alpha/2} \cdot (1.30)/\sqrt{5}$
$1.24 = t_{4,\alpha/2} \cdot (5828)$
$t_{4,\alpha/2} = 1.24/.5828 = 2.128$ [closest entry $= 2.132$], $\alpha/2 = .05$, $\alpha = .10$
The level of confidence is $1 - \alpha = 1 - .10 = .90 = 90\%$.

6-4 Determining Sample Size Required to Estimate μ

1. $E = 1.5$ and $\alpha = .02$
$n = [z_{.01} \cdot \sigma/E]^2$
$= [2.327 \cdot 15/1.5]^2$
$= 541.49$, rounded up to 542

3. $E = 2$ and $\alpha = .04$
$n = [z_{.02} \cdot \sigma/E]^2$
$= [2.05 \cdot 12.46/2]^2$
$= 163.11$, rounded up to 164

5. $E = 500$ and $\alpha = .05$
$n = [z_{.025} \cdot \sigma/E]^2$
$= [1.960 \cdot 6250/500]^2$
$= 600.25$, rounded up to 601

7. $E = .25$ and $\alpha = .04$
$$n = [z_{.02} \cdot \sigma/E]^2$$
$$= [2.05 \cdot 1.87/.25]^2$$
$$= 235.13, \text{ rounded up to } 236$$

9. estimated maximum age = 10 years; estimated minimum age = 0 years; $R = 10 - 0 = 10$
$E = .25$ and $\alpha = .04$; $\sigma \approx R/4 = 10/4 = 2.5$
$$n = [z_{.02} \cdot \sigma/E]^2$$
$$= [2.05 \cdot 2.5/.25]^2$$
$$= 420.25, \text{ rounded up to } 421$$

11. a. for the $n = 53$ Wednesdays, $s = .1346$
$E = .02$ and $\alpha = .02$
$$n = [z_{.01} \cdot \sigma/E]^2$$
$$= [2.327 \cdot .1346/.02]^2$$
$$= 245.26, \text{ rounded up to } 246$$

b. for the $n = 52$ Sundays, $s = .2000$
$E = .02$ and $\alpha = .02$
$$n = [z_{.01} \cdot \sigma/E]^2$$
$$= [2.327 \cdot .2000/.02]^2$$
$$= 541.49, \text{ rounded up to } 542$$

c. The results of parts (a) and (b) indicate that the required sample size can vary significantly according to the estimate used for σ. This illustrates the importance and necessity of planning for the worst possible scenario by using a large estimate for σ. It happens that Wednesday ($s = .1346$) was the least variable day. One might be tempted to use the most variable day, Saturday ($s = .2900$), to produce a conservative estimate of $n = 1139$. The best estimate, however, would probably be obtained by using all 365 days ($s = .2125$).

13. $E = 1.5$ and $\alpha = .02$
$N = 200$
$$n = [N\sigma^2(z_{.01})^2]/[(N-1)E^2 + \sigma^2(z_{.01})^2]$$
$$= [(200)(15)^2(2.327)^2]/[(199)(1.5)^2 + (15)^2(2.327)^2]$$
$$= [243671.8]/[1666.109]$$
$$= 146.25, \text{ rounded up to } 147$$

6-5 Estimating a Population Proportion

IMPORTANT NOTE: When calculating confidence intervals using the formula
$$\hat{p} \pm E$$
$$\hat{p} \pm z_{\alpha/2}\sqrt{\hat{p}\hat{q}/n}$$
do not round off in the middle of the problem. This may be accomplished conveniently on most calculators having a memory as follows.

(1) Calculate $\hat{p} = x/n$ and STORE the value
(2) Calculate E as 1 - RECALL = * RECALL = ÷ n = √ * $z_{\alpha/2}$ =
(3) With the value of E showing on the display, the upper confidence limit is calculated by + RECALL.
(4) With the value of the upper confidence limit showing on the display, the lower confidence limit is calculated by - RECALL ± + RECALL

You must become familiar with your own calculator. [Do your homework using the same type of calculator you will be using for the exams.] The above procedure works on most calculators; make certain you understand why it works and verify whether it works on your calculator. If it does not seem to work on your calculator, or if your calculator has more than one memory so that you can STORE both $\hat{p}$ and E at the same time, ask your instructor for assistance.

NOTE: It should be true that $0 \le \hat{p} \le 1$ and that $E \le .5$ [usually, <u>much</u> less than .5]. If such is not the case, an error has been made.

1. let L = the lower confidence limit; U = the upper confidence limit
   $\hat{p} = (L + U)/2$
   $= (.800 + .840)/2 = 1.640/2 = .820$
   $E = (U - L)/2$
   $= (.840 - .800)/2 = .040/2 = .020$

3. let L = the lower confidence limit; U = the upper confidence limit
   $\hat{p} = (L + U)/2$
   $= (.432 + .455)/2 = .4435$
   $E = (U - L)/2$
   $= (.455 - .432)/2 = .023/2 = .0115$

5. $\alpha = .05$ and $\hat{p} = x/n = 400/500 = .80$
   $E = z_{.025}\sqrt{\hat{p}\hat{q}/n}$
   $= 1.960\sqrt{(.80)(.20)/500} = .0351$

7. $\alpha = .02$ and $\hat{p} = x/n = 237/1068 = .222$
   $E = z_{.01}\sqrt{\hat{p}\hat{q}/n}$
   $= 2.327\sqrt{(.222)(.778)/1068} = .0296$

9. $\alpha = .05$ and $\hat{p} = x/n = .35$
   $E = z_{.025}\sqrt{\hat{p}\hat{q}/n}$
   $= 1.960\sqrt{(.35)(.65)/200} = .0661$

11. $\alpha = .01$ and $\hat{p} = x/n = 150/750 = .200$
    $\hat{p} \pm z_{.001}\sqrt{\hat{p}\hat{q}/n}$
    $.200 \pm 2.575\sqrt{(.200)(.800)/750}$
    $.200 \pm .038$
    $.162 < p < .238$

13. $\alpha = .05$ and $\hat{p} = x/n = 222/1357 = .164$
    $\hat{p} \pm z_{.025}\sqrt{\hat{p}\hat{q}/n}$
    $.164 \pm 1.960\sqrt{(.164)(.836)/1357}$
    $.164 \pm .020$
    $.144 < p < .183$

15. $\hat{p}$ unknown, use $\hat{p} = .5$
    $n = [(z_{.005})^2\hat{p}\hat{q}]/E^2$
    $= [(2.575)^2(.5)(.5)]/(.025)^2$
    $= 2652.25$, rounded up to 2653

17. $\hat{p} = .15$
    $n = [(z_{.025})^2\hat{p}\hat{q}]/E^2$
    $= [(1.960)^2(.15)(.82)]/(.03)^2$
    $= 544.23$, rounded up to 545

19. a. We are 99% certain that the interval from 93.053% to 94.926% contains the true percentage of U.S. households having telephones.

b. Based on the preceding result, pollsters should not be too concerned about results from surveys conducted by telephone if (1) proper techniques are used and (2) issues that would elicit differing responses from phone and non-phone households are not at stake. Proper techniques involve follow up of non-responses (due to work schedules, etc.), making certain the appropriate person in the household (and not just whoever answers the phone, etc.) is responding, etc. Issues involving agricultural policies, for example, would miss the opinions of the Amish (who are primarily involved in farming and do not have phones for religious reasons). In general, however, one can probably assume that those households with phones would be the households most likely to buy, use or be familiar with the subject of the survey.

21. a. $\hat{p} = x/n = 111/1233 = .090 = 9.0\%$
    b. $\hat{p} \pm z_{.025}\sqrt{\hat{p}\hat{q}/n}$
       $.090 \pm 1.960\sqrt{(.090)(.910)/1233}$
       $.090 \pm .016$
       $.074 < p < .106$
       $7.4\% < p < 10.6\%$

23. a. $\alpha = .05$ and $\hat{p} = x/n = 5.2\% = .052$
       $\hat{p} \pm z_{.025}\sqrt{\hat{p}\hat{q}/n}$
       $.052 \pm 1.960\sqrt{(.052)(.948)/8411}$
       $.052 \pm .005$
       $.047 < p < .057$
       $4.7\% < p < 5.7\%$

    NOTE: Since $\hat{p} = x/n = 5.2\%$ was given to only the nearest tenth of a percent, the usual 3 significant digit confidence interval is not possible. Additional accuracy cannot by obtained without knowing the true value of x -- and any x from 434 to 441 inclusive would give an x/8411 that rounds to .052.

    b. Yes; since the intervals are far from overlapping, the results are substantially different from those in part (a). In the presence of an explosion or fire on the ground, the likelihood of a pilot fatality is significantly increased.

25. a. $\alpha = .01$ and $\hat{p} = x/n = 7/221 = .032$
       NOTE: Since $\hat{p} = x/n = 3.2\%$ was given, we may assume $x = \hat{p}\cdot n = .032\cdot221 = 7$. No other integer value x/221 rounds to .032.
       $\hat{p} \pm z_{.005}\sqrt{\hat{p}\hat{q}/n}$
       $.032 \pm 2.575\sqrt{(.032)(.968)/221}$
       $.032 \pm .030$
       $.00134 < p < .0620$
       $0.134\% < p < 6.20\%$

    b. Since 1.8% is within the interval in part (a), it cannot be ruled out as the correct value for the true proportion of Ziac users that experience dizziness. The test does not provide evidence that Ziac users experience any more dizziness than non-users -- i.e., the test does not provide evidence that dizziness is an adverse reaction to Ziac.

27. a. $\hat{p} = .82$
       $n = [(z_{.01})^2\hat{p}\hat{q}]/E^2$
         $= [(2.327)^2(.82)(.18)]/(.05)^2$
         $= 319.70$, rounded up to 320
    b. $\hat{p}$ unknown, use $\hat{p} = .5$
       $n = [(z_{.01})^2\hat{p}\hat{q}]/E^2$
         $= [(2.327)^2(.5)(.5)]/(.05)^2$
         $= 541.49$, rounded up to 542

29. a. $\hat{p} = .86$
$n = [(z_{.03})^2\hat{p}\hat{q}]/E^2$
$= [(1.88)^2(.86)(.14)]/(.03)^2$
$= 472.82$, rounded up to 473
b. $\hat{p}$ unknown, use $\hat{p} = .5$
$n = [(z_{.03})^2\hat{p}\hat{q}]/E^2$
$= [(1.88)^2(.5)(.5)]/(.03)^2$
$= 981.78$, rounded up to 982

31. a. $\alpha = .10$ and $\hat{p} = x/n = 7/80 = .0875$
$\hat{p} \pm z_{.01}\sqrt{\hat{p}\hat{q}/n}$
$.0875 \pm 1.645\sqrt{(.0875)(.9125)/80}$
$.0875 \pm .0520$
$.036 < p < .139$
b. $\hat{p} = .0875$
$n = [(z_{.02})^2\hat{p}\hat{q}]/E^2$
$= [(2.05)^2(.0875)(.9125)]/(.03)^2$
$= 372.83$, rounded up to 373
c. Yes; since $0.25\% = .0025$ is not in within the confidence interval in part (a), we may safely conclude that women have a lower rate of red/green color blindness than men do.

32. a. $\alpha = .03$ and $\hat{p} = x/n = 22\% = .22$
$\hat{p} \pm z_{.015}\sqrt{\hat{p}\hat{q}/n}$
$.22 \pm 2.17\sqrt{(.22)(.78)/4000}$
$.22 \pm .01$
$.21 < p < .23$
NOTE: Since $\hat{p} = x/n = 22\%$ was given to only the nearest whole percent, the usual 3 significant digit confidence interval is not possible. Additional accuracy cannot by obtained without knowing the true value of x -- and any x between 860 and 900 would give an x/4000 that rounds to .22.
b. $\hat{p}$ unknown, use $\hat{p} = .5$
$n = [(z_{.005})^2\hat{p}\hat{q}]/E^2$
$= [(2.575)^2(.5)(.5)]/(.005)^2$
$= 66306.25$, rounded up to 66307
c. Yes; since $11\% = .11$ is not in within the confidence interval in part (a), we may safely conclude that 60 Minutes had a greater share of the audience than the wrestling special did. No; exposing professional wrestling was probably not necessary.

33. $\alpha = .05$ and $\hat{p} = x/n = 21/100 = .210$
$\hat{p} \pm z_{.025}\sqrt{\hat{p}\hat{q}/n}$
$.210 \pm 1.960\sqrt{(.210)(.790)/100}$
$.210 \pm .080$
$.130 < p < .290$
$13.0\% < p < 29.0\%$
Yes; this result is consistent with the 20% rate reported by the candy maker.

35. $\hat{p}$ unknown, use $\hat{p} = .5$
$n = [N\hat{p}\hat{q}(z_{.005})^2]/[\hat{p}\hat{q}(z_{.005})^2 + (N-1)E^2]$
$= [(5000)(.5)(.5)(2.575)^2]/[(.5)(.5)(2.575)^2 + (4999)(.005)^2]$
$= [8288.281]/[1.782631]$
$= 4649.46$, rounded up to 4650

6-6 Estimating a Population Variance

1. $\chi_L^2 = \chi_{29,.975}^2 = 16.047$; $\chi_R^2 = \chi_{29,.025}^2 = 45.722$

3. $\chi_L^2 = \chi_{49,.995}^2 = 27.991$; $\chi_R^2 = \chi_{49,.005}^2 = 79.490$
   NOTE: Use df = 50, the closest entry.

5. $\qquad (n-1)s^2/\chi_{14,.025}^2 < \sigma^2 < (n-1)s^2/\chi_{14,.975}^2$
   $(14)(108)^2/26.119 < \sigma^2 < (14)(108)^2/5.629$
   $6252 < \sigma^2 < 29009$
   $79 < \sigma < 170$

7. $\qquad (n-1)s^2/\chi_{24,.05}^2 < \sigma^2 < (n-1)s^2/\chi_{24,.95}^2$
   $(24)(12)^2/36.415 < \sigma^2 < (24)(12)^2/13.848$
   $94.9 < \sigma^2 < 249.6$
   $10 < \sigma < 16$

9. From the upper right section of Table 6-3, n = 767.

11. From the lower left section of Table 6-3, n = 1,401.

NOTE: When raw scores are available, $\bar{x}$ and s should be calculated as the primary descriptive statistics -- but use the unrounded value of s^2 in the confidence interval formula. In addition, always make certain that the confidence interval for σ includes the calculated value of s.

13. summary information
n = 18	$\bar{x}$ = 3787.0
Σx = 68,166	s^2 = 3066.82
Σx^2 = 258,196,778	s = 55.4

    $\qquad (n-1)s^2/\chi_{17,.005}^2 < \sigma^2 < (n-1)s^2/\chi_{17,.995}^2$
    $(17)(3066.82)/35.718 < \sigma^2 < (17)(3066.82)/5.697$
    $1459.65 < \sigma^2 < 9151.47$
    $38.2 < \sigma < 95.7$ (mL)

    No; the interval indicates 99% confidence that σ > 30.

15. a. s $\approx (x_n - x_1)/4 = (1.033 - .856)/4 = .177/4 = .04425$
    b. summary information
n = 33	$\bar{x}$ = .9128
Σx = 30.123	s^2 = .0015616
Σx^2 = 27.546793	s = .0395

    $\qquad (n-1)s^2/\chi_{32,.01}^2 < \sigma^2 < (n-1)s^2/\chi_{32,.99}^2$
    $(32)(.0015616)/50.892 < \sigma^2 < (32)(.0015616)/14.954$
    $.0009819 < \sigma^2 < .0033416$
    $.0313 < \sigma < .0578$ (grams)

    NOTE: The χ^2 values used are those for 30 df, which is the closest entry from Table A-4. Since df = n-1 is a part of the confidence interval formula, and since the expected value of the χ^2 statistic is n-1, some instructors and text books recommend using in the formula a value for n-1 that agrees with the df of the tabled χ^2
    c. Yes; .04425 is within the confidence interval.

17. a. summary information

$n = 10$ $\bar{x} = 7.15$
$\Sigma x = 71.5$ $s^2 = .2272$
$\Sigma x^2 = 513.27$ $s = .48$

$(n-1)s^2/\chi^2_{9,.025} < \sigma^2 < (n-1)s^2/\chi^2_{9,.975}$
$(9)(.2272)/19.023 < \sigma^2 < (9)(.2272)/2.700$
$.1075 < \sigma^2 < .7573$
$.33 < \sigma < .87$ (minutes)

b. summary information

$n = 10$ $\bar{x} = 7.15$
$\Sigma x = 71.5$ $s^2 = 3.3183$
$\Sigma x^2 = 541.09$ $s = 1.82$

$(n-1)s^2/\chi^2_{9,.025} < \sigma^2 < (n-1)s^2/\chi^2_{9,.975}$
$(9)(3.3183)/19.023 < \sigma^2 < (9)(3.3183)/2.700$
$1.5699 < \sigma^2 < 11.0610$
$1.25 < \sigma < 3.33$ (minutes)

c. Yes, there is a difference. The multiple line system exhibits more variability among the waiting times. The greater consistency among the single line waiting times seems fairer to the customers and more professional.

19. a. summary information

$n = 25$ $\bar{x} = 78.898$
$\Sigma x = 1972.45$ $s^2 = 192.373$
$\Sigma x^2 = 160239.3228$ $s = 13.87$

$(n-1)s^2/\chi^2_{24,.01} < \sigma^2 < (n-1)s^2/\chi^2_{24,.99}$

$(24)(192.373)/42.980 < \sigma^2 < (24)(192.373)/10.856$
$107.421 < \sigma^2 < 425.291$
$10.364 < \sigma < 20.623$ (dollars)

b. summary information

$n = 15$ $\bar{x} = 42.150$
$\Sigma x = 632.25$ $s^2 = 249.279$
$\Sigma x^2 = 30139.2424$ $s = 15.79$

$(n-1)s^2/\chi^2_{14,.01} < \sigma^2 < (n-1)s^2/\chi^2_{14,.99}$

$(14)(249.279)/29.141 < \sigma^2 < (14)(249.279)/4.660$
$119.759 < \sigma^2 < 748.907$
$10.943 < \sigma < 27.366$ (dollars)

c. There does not appear to be a significant difference in the variability of the two groups.

21. a. The given interval $2.8 < \sigma < 6.0$
$7.84 < \sigma^2 < 36.00$
and the usual calculations $(n-1)s^2/\chi^2_{19,\alpha/2} < \sigma^2 < (n-1)s^2/\chi^2_{19,1-\alpha/2}$
$(19)(3.8)^2/\chi^2_{19,\alpha/2} < \sigma^2 < (19)(3.8)^2/\chi^2_{19,1-\alpha/2}$
$274.36/\chi^2_{19,\alpha/2} < \sigma^2 < 274.36/\chi^2_{19,1-\alpha/2}$

imply that $7.84 = 274.37/\chi^2_{19,\alpha/2}$ and $36.00 = 274.36/\chi^2_{19,1-\alpha/2}$

$\chi^2_{19,\alpha/2} = 274.36/7.84$ $\chi^2_{19,1-\alpha/2} = 274.36/36.00$
$= 34.99$ $= 7.62$

The closest entries in Table A-4 are $\chi^2_{19,\alpha/2} = 34.805$ and $\chi^2_{19,1-\alpha/2} = 7.633$
which imply $\alpha/2 = .01$ $1 - \alpha/2 = .99$
$\alpha = .02$ $\alpha/2 = .01$
$\alpha = .02$

The level of confidence is therefore is $1-\alpha = 98\%$.

b. $(n-1)s^2/\chi^2_{11,.025} < \sigma^2 < (n-1)s^2/\chi^2_{11,.975}$

using the lower endpoint	OR	using the upper endpoint
$(11)s^2/21.920 = (19.1)^2$		$(11)s^2/3.816 = (45.8)^2$
$s^2 = 726.97$		$s^2 = 727.69$
$s = 27.0$		$s = 27.0$

Review Exercises

1. a. $n > 30$, use $\underline{z}$ (with s for σ)

   $\bar{x} \pm z_{.01} \cdot \sigma/\sqrt{n}$

   $4.4038 \pm 2.327 \cdot 2.8440/\sqrt{52}$

   $4.4038 \pm .9177$

   $3.5 < \mu < 5.3$

   b. $n > 30$, use $\underline{z}$ (with s for σ)

   $\bar{x} \pm z_{.01} \cdot \sigma/\sqrt{n}$

   $1.7308 \pm 2.327 \cdot 1.7502/\sqrt{52}$

   $1.7308 \pm .5648$

   $1.2 < \mu < 2.3$

   c. Since the intervals do not overlap we can be confident that the population values are truly different. The Dvorak ratings are significantly lower than the QWERTY ratings.

   d. The methods of Section 6-6 require that the populations have normal distributions. These population distributions are not normal -- they are truncated on the left at 0 and skewed to the right.

2. $\hat{p}$ unknown, use $\hat{p} = .5$

   $n = [(z_{.015})^2 \hat{p}\hat{q}]/E^2$

   $= [(2.17)^2(.5)(.5)]/(.02)^2$

   $= 2943.06$, rounded up to 2944

3. a. $E = 2$ and $\alpha = .03$

   $n = [z_{.015} \cdot \sigma/E]^2$

   $= [2.17 \cdot 15.7/2]^2$

   $= 290.17$, rounded up to 291

   b. The closest colleges are not necessarily representative of all colleges, and they may differ in ways that would affect the scores -- size of the classes, admission standard, types of majors offered, etc. A convenience sample is not an appropriate substitute for a random sample.

4. a. $n \leq 30$ and σ unknown, use t

   $\bar{x} \pm t_{24,.025} \cdot s/\sqrt{n}$

   $7.01 \pm 2.064 \cdot 3.741/\sqrt{25}$

   7.01 ± 1.54

   $5.47 < \mu < 8.55$ (years)

   b. $(n-1)s^2/\chi^2_{24,.025} < \sigma^2 < (n-1)s^2/\chi^2_{24,.975}$

   $(24)(3.74)^2/39.364 < \sigma^2 < (24)(3.74)^2/12.401$

   $8.528 < \sigma^2 < 27.071$

   $2.92 < \sigma < 5.20$ (years)

5. $\hat{p} = x/n = 61/94 = .649$
$\alpha = .05$
$\hat{p} \pm z_{.025}\sqrt{\hat{p}\hat{q}/n}$
$.649 \pm 1.960\sqrt{(.649)(.351)/94}$
$.649 \pm .096$
$.552 < p < .745$
$55.2\% < p < 74.5\%$
We are 95% certain that the interval from 55.2% to 74.5% contains the true proportion of automobile air bag fatalities that involved improper belting.

6. $\hat{p} = .93$
$n = [(z_{.01})^2\hat{p}\hat{q}]/E^2$
$= [(2.327)^2(.93)(.07)]/(.04)^2$
$= 220.32$, rounded up to 221

7. preliminary information: $n = 16$, $\Sigma x = 1162.6$, $\Sigma x^2 = 84577.34$
$\overline{x} = (\Sigma x)/n = 1162.6/16 = 72.66$ [do not round, store <u>all</u> the digits in the calculator]
$s^2 = [n(\Sigma x^2) - (\Sigma x)^2]/[n(n-1)]$
$= [16(84577.34) - (1162.6)^2]/[16(15)]$
$= 6.661$
$s = 2.581$ [do not round, store <u>all</u> the digits in the calculator]
$n \le 30$ and σ unknown, use t
$\overline{x} \pm t_{15,.025} \cdot s/\sqrt{n}$
$72.66 \pm 2.132 \cdot 2.581/\sqrt{16}$
72.66 ± 1.38
$71.29 < \mu < 74.04$ (cm)

8. $n > 30$, use z (with s for σ)
$\overline{x} \pm z_{.005} \cdot \sigma/\sqrt{n}$
$558 \pm 2.575 \cdot 139/\sqrt{67}$
558 ± 44
$514 < \mu < 602$

9. $E = 4$ and $\alpha = .05$
$n = [z_{.025} \cdot \sigma/E]^2$
$= [1.960 \cdot 41.0/4]^2$
$= 403.61$, rounded up to 404

10. $\hat{p} = .24$
$\alpha = .01$
$\hat{p} \pm z_{.005}\sqrt{\hat{p}\hat{q}/n}$
$.24 \pm 2.575\sqrt{(.24)(.76)/1998}$
$.24 \pm .02$
$.22 < p < .26$
$22\% < p < 26\%$
NOTE: Since $\hat{p}$ was given with only two decimal accuracy and the actual value of x was not given, the final answer is limited to two decimal accuracy. Any x from 470 to 489 inclusive rounds to 24% with varying digits for the next decimal point.

Cumulative Review Exercises

1. Begin by making a stem-and-leaf plot and calculating summary statistics.

```
10 | 5
11 |                      n = 9
11 | 599                  Σx = 1089
12 | 3                    Σx² = 132223
12 | 5788
```

a. $\bar{x} = (\Sigma x)/n = (1089)/9 = 121.0$ lbs

b. $\tilde{x} = 123\ 0$ lbs

c. $M = 119, 128$ (bi-modal)

d. m.r. $= (105 + 128)/2 = 116.5$ lbs

e. $R = 128 - 105 = 23$ lbs

f. $s^2 = [n(\Sigma x^2) - (\Sigma x)^2]/[n(n-1)]$
$= [9(132223) - (1089)^2]/[9(8)]$
$= 56.75$ lbs^2

g. $s = 7.5$ lbs

h. for $Q_1 = P_{25}$, $L = (25/100)(9) = 2.25$, round up to 3
$Q_1 = x_3 = 119$ lbs

i. for $Q_2 = P_{50}$, $L = (50/100)(9) = 4.50$, round up to 5
$Q_2 = x_5 = 123$ lbs

j. for $Q_3 = P_{75}$, $L = (75/100)(9) = 6.75$, round up to 7
$Q_3 = x_7 = 127$ lbs

k. ratio, since differences are
consistent and there is a meaningful zero

l. The boxplot is at the right.

m. $n \le 30$ and σ unknown, use t
$\bar{x} \pm t_{8,.005} \cdot s/\sqrt{n}$
$121.0 \pm 3.355 \cdot 7.5/\sqrt{9}$
121.0 ± 8.4
$112.6 < \mu < 129.4$ (lbs)

n. Since both the stem-and-leaf plot and the boxplot indicate that the weights do not appear to come from a normal population, the methods of Section 6-6 are not appropriate. It appears the distribution of the weights of supermodels is truncated at some upper limit and skewed to the left. Even though a confidence interval for σ should not be constructed, the following interval is given as a review of the technique.
$(n-1)s^2/\chi^2_{8,.005} < \sigma^2 < (n-1)s^2/\chi^2_{8,.995}$
$(8)(56.75)/21.955 < \sigma^2 < (8)(56.75)/1.344$
$20.68 < \sigma^2 < 337.80$
$4.5 < \sigma < 18.4$ (lbs)

o. $E = 2$ and $\alpha = .01$
$n = [z_{.005} \cdot \sigma/E]^2$
$= [2.575 \cdot 7.5/2]^2$
$= 94.07$, rounded up to 95

p. For the general female population, an unusually low weight would be one below 85 lbs (i.e., more than 2σ below μ). Individually, none of the supermodels has an unusually low weight. As a group, however, they are each well below the general population mean weight and appear to weigh substantially less than the general population -- as evidenced by their mean weight as given by the point estimate in part (a) and the interval estimate in part (m).

2. a. binomial: n = 200 and p = .25
    the normal approximation is appropriate since
       np = 200(.25) = 50 ≥ 5
       n(1-p) = 200(.75) = 150 ≥ 5
    use μ = np = 200(.25) = 50
       σ = √n(p)(1-p) = √200(.25)(.75) = 6.214
    P(x ≥ 65)
       = P_c(x > 64.5)
       = P(z > 2.37)
       = .5000 - .4911
       = .0089

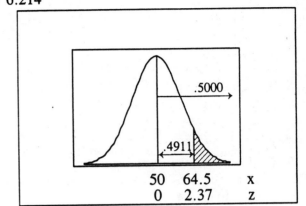

.5000

.4911

| 50 | 64.5 | x |
| 0 | 2.37 | z |

   b. p̂ = 65/200 = .325
       α = .05
       p̂ ± $z_{.025}$√p̂q̂/n
       .325 ± 1.960√(.325)(.675)/200
       .325 ± .065
       .260 < p < .390
   c. No; the expert's value of .25 does not seem correct for two reasons.
       From part (a): if the expert is correct, then the probability of getting the sample
             obtained is very small -- less than 1%.
       From part (b): according to the sample obtained, there is 95% confidence that the
             interval from .26 to .39 includes the true value -- that is, there is 95%
             confidence that the value is not .25.

Chapter 7

Hypothesis Testing

7-2 Fundamentals of Hypothesis testing

1. Conclude that the coin is not fair. Since the occurrence of 27 heads in 30 tosses would be a rare event if the coin were fair, we reject the claim that it is fair.

3. Do not conclude that women who eat blue M&M's have a better chance of having a baby boy. Since the occurrence of 27 boys in 50 births is not a rare event if the candy has no effect, it is not significant evidence to support the stated claim.

5. claim: Lisa's cars are stolen at the "normal" rate.
   conclusion: Lisa's cars are stolen at a higher rate than normal, which may be for any number of reasons -- the models she buys or places she drives, her carefulness (or lack thereof), etc..

7. claim: The experimental vaccine is effective against Lyme disease.
   conclusion: The evidence is not convincing. The difference between the 2.2% rate of infection for those receiving the placebo and the 1.8% rate for those receiving the vaccine is not significant -- such a difference in the sample rates could easily occur by chance when the population rates are identical and the vaccine has no effect.

9. original statement: $\mu > 30$
   competing idea: $\mu \leq 30$ (contains the equality; must be H_o)
       $H_o: \mu \leq 30$
       $H_1: \mu > 30$

11. original statement: $p < .5$
    competing idea: $p \geq .5$ (contains the equality; must be H_o)
        $H_o: p \geq .5$
        $H_1: p < .5$

13. original statement: $\sigma < 2.8$
    competing idea: $\sigma \geq 2.8$ (contains the equality; must be H_o)
        $H_o: \sigma \geq 2.8$
        $H_1: \sigma < 2.8$

15. original statement: $\mu \geq 12$ (contains the equality; must be H_o)
    competing idea: $\mu < 12$
        $H_o: \mu \geq 12$
        $H_1: \mu < 12$

17. two-tailed test; place $\alpha/2$ in each tail
    use A = .5000 - $\alpha/2$ = .5000 -.0250 = .4750
    critical values are $\pm z_{\alpha/2}$ = $\pm z_{.025}$ = ± 1.960

19. right-tailed test; place α in the upper tail
    use A = .5000 - α = .5000 - .0200 = .4800 [closest entry = .4798]
    critical value is $+z_{\alpha}$ = $+z_{.02}$ = $+2.05$

21. two-tailed test; place $\alpha/2$ in each tail
     use A = .5000 - $\alpha/2$ = .5000 -.0200 = .4800 [closest entry = .4798]
     critical values are $\pm z_{\alpha/2}$ = $\pm z_{.02}$ = ± 2.05

23. right-tailed test; place α in the upper tail
     use A = .5000 - α = .5000 -.0600 = .4400 [half-way between .4394 and .4406]
     critical value is $+z_{\alpha}$ = $+z_{.06}$ = $+1.555$

25. $z_{\bar{x}}$ = $(\bar{x} - \mu)/(\sigma/\sqrt{n})$ [use s to estimate σ]
        = $(.83 - .21)/(.24/\sqrt{32})$
        = $(.62)/(.0424)$
        = 14.614

27. $z_{\bar{x}}$ = $(\bar{x} - \mu)/(\sigma/\sqrt{n})$ [use s to estimate σ]
        = $(.83 - 1.39)/(.16/\sqrt{123})$
        = $(-.56)/(.0144)$
        = - 38.817

29. original claim: μ = 60 (contains the equality; must be H_o)
     competing idea: $\mu \neq 60$
        H_o: μ = 60
        H_1: $\mu \neq 60$
     initial conclusion: Reject H_o.
     final conclusion: There is sufficient evidence to reject the claim that women have a mean
        height of 60.0 inches.

31. original claim: $\mu < 12$
     competing idea: $\mu \geq 12$ (contains the equality; must be H_o)
        H_o: $\mu \geq 12$
        H_1: $\mu < 12$
     initial conclusion: Fail to reject H_o.
     final conclusion: There is not sufficient evidence to support the claim that the mean amount
        of Coke in 12 ounce cans is actually less than 12 ounces.

33. original claim: μ = 12 (contains the equality; must be H_o)
     competing idea: $\mu \neq 12$
        H_o: μ = 12
        H_1: $\mu \neq 12$
     type I error: Reject the original claim that μ = 12 when μ = 12 is actually true.
     type II error: Fail to reject the original claim that μ = 12 when $\mu \neq 12$ is actually true.

35. original claim: $\mu < 100$
     competing idea: $\mu \geq 100$ (contains the equality; must be H_o)
        H_o: $\mu \geq 100$
        H_1: $\mu < 100$
     type I error: Reject the null hypothesis, and conclude the original claim that $\mu < 100$ is
        true when $\mu \geq 100$ is actually true.
     type II error: Fail to reject the null hypothesis, and fail to conclude the original claim that
        $\mu < 100$ is true when $\mu < 100$ is actually true.

37. The statistical test is based on rejecting results that represent an extreme departure from
     H_o. When testing the null hypothesis that $\mu \geq 100$, it is necessary to conduct the test
     based on the assumption that μ = 100 instead of $\mu \geq 100$ for two reasons:
     (1) The test assumption must be specific so that we can decide exactly what we expect and
        measure how much the results depart from that expectation. While μ = 100 is
        specific and we know to expect $\bar{x}$'s normally distributed around 100, $\mu \geq 100$ is not

specific.
(2) Testing $\mu = 100$ is sufficient. If there is enough evidence to reject $\mu = 100$ to conclude that $\mu < 100$, there is certainly enough evidence to reject any other specific value greater than 100.

39. Mathematically, in order for α to equal 0 the magnitude of the critical value would have to be infinite. Practically, the only way never to make a type I error is to always fail to reject H_o. From either perspective, the only way to achieve $\alpha = 0$ is to never reject H_o no matter how extreme the sample data might be.

7-3 Testing a Claim about a Mean: Large Samples

1. H_o: $\mu = 100$ the test is two-tailed
   H_1: $\mu \neq 100$
   P-value = P(getting a sample result at least as extreme as the one obtained)
   $\qquad = 2 \cdot P(z < -1.03)$
   $\qquad = 2 \cdot (.5000 - .3485) = 2 \cdot (.1515) = .3030$

3. H_o: $\mu \geq 500$ the test is left-tailed
   H_1: $\mu < 500$
   P-value = P(getting a sample result at least as extreme as the one obtained)
   $\qquad = P(z < -12.67)$
   $\qquad = .5000 - .4999 = .0001$

5. original claim: $\mu = 75$ [n > 30, use z (with s for σ)]
   H_o: $\mu = 75$
   H_1: $\mu \neq 75$
   $\alpha = .05$
   a. test statistic: $z_{\bar{x}} = (\bar{x} - \mu)/\sigma_{\bar{x}}$
   $\qquad\qquad\qquad = (77 - 75)/(15/\sqrt{500})$
   $\qquad\qquad\qquad = 2/.671$
   $\qquad\qquad\qquad = 2.981$
   b. critical values: $z = \pm z_{.025} = \pm 1.960$
   c. P-value: $2 \cdot P(z > 2.98) = 2 \cdot (.5000 - .4986)$
   $\qquad\qquad\qquad\qquad\quad = 2 \cdot (.0014)$
   $\qquad\qquad\qquad\qquad\quad = .0028$
   d. final conclusion: Reject H_o; there is sufficient evidence to reject the claim that $\mu = 75$ and conclude that $\mu \neq 75$ (in fact, that $\mu > 75$).

7. original claim: $\mu > 2.25$ [n > 30, use z (with s for σ)]
   H_o: $\mu \leq 2.50$
   H_1: $\mu > 2.50$
   $\alpha = .02$
   a. test statistic: $z_{\bar{x}} = (\bar{x} - \mu)/\sigma_{\bar{x}}$
   $\qquad\qquad\qquad = (2.35 - 2.25)/(.80/\sqrt{150})$
   $\qquad\qquad\qquad = .10/.0653$
   $\qquad\qquad\qquad = 1.53$
   b. critical value: $z = +z_{.02} = +2.05$
   c. P-value: $P(z > 1.53) = .5000 - .4370$
   $\qquad\qquad\qquad\qquad = .0630$
   d. final conclusion: Do not reject H_o; there is not sufficient evidence to support the claim that $\mu > 2.25$.

9. original claim: $\mu < 98.6$ [n > 30, use z (with s for σ)]
   H_o: $\mu \geq 98.6$
   H_1: $\mu < 98.6$
   $\alpha = .01$
   C.R. $z < -z_{.01} = -2.327$
   calculations:
       $z_{\bar{x}} = (\bar{x} - \mu)/\sigma_{\bar{x}}$
          $= (98.20 - 98.6)/(.62/\sqrt{106})$
          $= -.40/.0602 = -6.642$

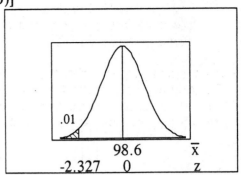

conclusion:
   Reject H_o; there is sufficient evidence to conclude that $\mu < 98.6$.

11. original claim: $\mu \neq 92.84$ [n > 30, use z (with s for σ)]
   H_o: $\mu = 92.84$
   H_1: $\mu \neq 92.84$
   $\alpha = .05$
   C.R. $z < -z_{.025} = -1.960$
       $z > z_{.025} = 1.960$
   calculations:
       $z_{\bar{x}} = (\bar{x} - \mu)/\sigma_{\bar{x}}$
          $= (92.67 - 92.84)/(1.79/\sqrt{40})$
          $= -.17/.283 = -.601$

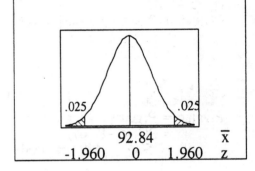

conclusion:
   Do not reject H_o; there is not sufficient evidence to conclude that $\mu \neq 92.84$.
No; the new balls do not appear to be significantly different.

13. original claim: $\mu = 600$ [n > 30, use z (with s for σ)]
   H_o: $\mu = 600$
   H_1: $\mu \neq 600$
   $\alpha = .01$
   C.R. $z < -z_{.005} = -2.575$
       $z > z_{.005} = 2.575$
   calculations:
       $z_{\bar{x}} = (\bar{x} - \mu)/\sigma_{\bar{x}}$
          $= (593 - 600)/(21/\sqrt{40})$
          $= -7/3.320 = -2.108$

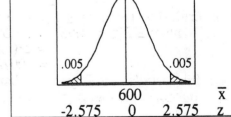

conclusion:
   Do not reject H_o; there is not sufficient evidence to reject the claim that $\mu = 600$.
While there is not enough evidence to reject the claim that $\mu = 600$ at the .01 level of significance, the results are certainly not evidence that $\mu = 600$ is true. As the error of 7 mg is only about 1% of the advertised amount, however, the product is probably not dangerous or unacceptable to use and may well be within acceptable limits of quality control. Based on this information alone, there is no reason not to buy this medicine.

15. original claim: $\mu > 0$ [$n > 30$, use z (with s for σ)]
    H_o: $\mu \le 0$
    H_1: $\mu > 0$
    $\alpha = .05$
    calculations:
    $z_{\bar{x}} = (\bar{x} - \mu)/\sigma_{\bar{x}}$
    $\phantom{z_{\bar{x}}} = (.6 - 0)/(3.8/\sqrt{75})$
    $\phantom{z_{\bar{x}}} = .6/.439 = 1.367$

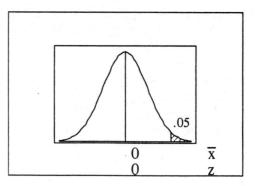

    P-value = P(z > 1.37) = .5000 - .4147 = .0853
conclusion:
    Do not reject H_o; there is not sufficient evidence to conclude that $\mu > 0$.
No; people should not take the course. Even if the mean increase of .6 had been statistically significant, it would not be of practical significance.

17. original claim: $\mu \ne 69.5$ [$n > 30$, use z (with s for σ)]
    H_o: $\mu = 69.5$
    H_1: $\mu \ne 69.5$
    $\alpha = .05$
    calculations:
    $z_{\bar{x}} = (\bar{x} - \mu)/\sigma_{\bar{x}}$
    $\phantom{z_{\bar{x}}} = (73.4 - 69.5)/(8.7/\sqrt{35})$
    $\phantom{z_{\bar{x}}} = 3.9/1.471 = 2.652$

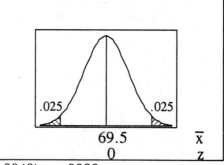

    P-value = 2·P(z > 2.651) = 2·(.5000 - .4960) = 2·(.0040) = .0080
conclusion:
    Reject H_o; there is sufficient evidence to conclude that $\mu \ne 69.5$ (in fact, that $\mu > 69.5$).

19. original claim: $\mu = 5.670$ [$n > 30$, use z (with s for σ)]
    H_o: $\mu = 5.670$
    H_1: $\mu \ne 5.670$
    $\alpha = .01$
    calculations:
    $z_{\bar{x}} = (\bar{x} - \mu)/\sigma_{\bar{x}}$
    $\phantom{z_{\bar{x}}} = (5.622 - 5.670)/(.068/\sqrt{50})$
    $\phantom{z_{\bar{x}}} = -.048/.00962 = -4.991$

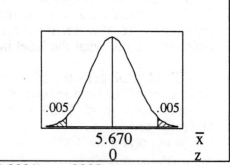

    P-value = 2·P(z < -4.99) = 2·(.5000 - .4999) = 2·(.0001) = .0002
conclusion:
    Reject H_o; there is sufficient evidence to conclude that $\mu \ne 5.670$ (in fact, $\mu < 5.670$).
That the mean weight of worn quarters found in circulation is less than 5.670 is not evidence that the mean weight of uncirculated quarters straight from the mint is different from 5.670.

21. original claim: $\mu = 3.5$ [$n > 30$, use z (with s for σ)]
    summary statistics: $n = 70$, $\Sigma x = 251.023$, $\Sigma x^2 = 900.557369$, $\bar{x} = 3.5860$, $s = .07403$
    H_o: $\mu = 3.5$
    H_1: $\mu \neq 3.5$
    $\alpha = .05$ [assumed]
    C.R. $z < -z_{.025} = -1.960$
    $\quad\quad z > z_{.025} = 1.960$
    calculations:

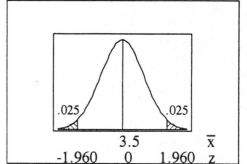

$$z_{\bar{x}} = (\bar{x} - \mu)/\sigma_{\bar{x}}$$
$$= (3.5860 - 3.5)/(.07403/\sqrt{70})$$
$$= .0860/.008848 = 9.725$$

    conclusion:
        Reject H_o; there is sufficient evidence to reject the claim that $\mu = 3.5$ and conclude that
        $\mu \neq 3.5$ (in fact, that $\mu > 3.5$).
    P-value $= 2 \cdot P(z > 9.72) = 2 \cdot (.5000 - .4999) = 2 \cdot (.0001) = .0002$

23. original claim: $\mu > 12$ [$n > 30$, use z (with s for σ)]
    summary statistics: $n = 36$, $\Sigma x = 442.5$, $\Sigma x^2 = 5439.35$, $\bar{x} = 12.29$, $s = .0906$
    H_o: $\mu \leq 12$
    H_1: $\mu > 12$
    $\alpha = .05$ [assumed]
    C.R. $z > z_{.05} = 1.645$
    calculations:

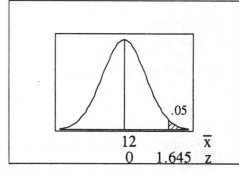

$$z_{\bar{x}} = (\bar{x} - \mu)/\sigma_{\bar{x}}$$
$$= (12.29 - 12)/(.0906/\sqrt{36})$$
$$= .2917/.0151 = 19.309$$

    conclusion:
        Reject H_o; there is sufficient evidence to conclude that $\mu > 12$.
    P-value $= P(z > 19.31) = .5000 - .4999 = .0001$
    Yes; If the company were to cut back slightly on the volume per can and keep the price the
    same, Pepsi could probably be supplied in a way that is more profitable to the company.
    But it is probably true that (1) the company is willing to continue overfilling to minimize
    the occurrence of underfilling and (2) the company bases its cost on what it actually
    supplies and not what the label indicates.

25. NOTE: Minitab calls the calculated statistic T instead of Z because it uses the notation and
    techniques of the next section, 7-4.
    original claim: $\mu < .9085$ [$n > 30$, use z (with s for σ)]
    H_o: $\mu \geq .9085$
    H_1: $\mu < .9085$
    $\alpha = .05$ [assumed]
    C.R. $z < -z_{.05} = -1.645$
    calculations:

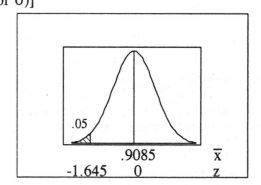

$$z_{\bar{x}} = (\bar{x} - \mu)/\sigma_{\bar{x}}$$
$$= (.91282 - .9085)/(.03952/\sqrt{33})$$
$$= .00432/.00688 = .63 \text{ [from Minitab]}$$

    conclusion:
        Do not reject H_o; there is not sufficient evidence to conclude that $\mu < .9085$.
    P-value $= P(z < .63) = .73$ [from Minitab]
    No; consumers are not being cheated.

27. original claim: $\mu = 98.6$ [n > 30, use z (with s for σ)]
H_o: $\mu = 98.6$
H_1: $\mu \neq 98.6$
$\alpha = .05$ [assumed]
C.R. z < $z_{.025}$ = -1.960
    z > $z_{.025}$ = 1.960
calculations:
$z_{\bar{x}} = (\bar{x} - \mu)/\sigma_{\bar{x}}$
$= (98.1263 - 98.6)/(.7554/\sqrt{38})$
$= -3.865$ [from TI-83 Plus]

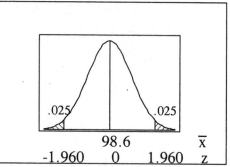

conclusion:
Reject H_o; there is sufficient evidence to reject the claim that $\mu = 98.6$ and to conclude that $\mu \neq 98.6$ (in fact, that $\mu < 98.6$).
P-value = 2·P(z < -3.865) = 1.109E-4 = .0001109 [from TI-83 Plus]
Yes; these results are consistent with those found in the example. Yes; the large number of missing values could be a problem -- not because of their quantity (the remaining sample size is large enough), but because they represent those unable to give an early morning reading. It is possible that those who were unable to be present at 8 am were those who were still in bed, and the desire to stay in bed later may conceivably be related to body temperature -- meaning that those who showed up were a biased sample.

29. original claim: $\mu = 100$ [n > 30, use z (with s for σ)]
H_o: $\mu = 100$
H_1: $\mu \neq 100$
$\alpha = .01$
C.R. z < $-z_{.005}$ = -2.575
    z > $z_{.005}$ = 2.575
calculations:
$z_{\bar{x}} = (\bar{x} - \mu)/\sigma_{\bar{x}}$
$2.575 < (103.6 - 100)/(s/\sqrt{62})$
$s < (103.6 - 100)/(2.575/\sqrt{62})$
$s < 3.6/.327$
$s < 11.008$

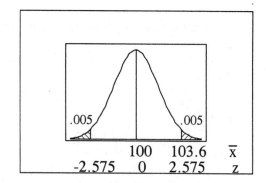

The largest possible standard deviation is 11.0.
NOTE: There is no conclusion because this exercise is not asking for the completion of a test of hypotheses.

31. original claim: $\mu = 98.6$ [n > 30, use z (with s for σ)]
H_o: $\mu = 98.6$
H_1: $\mu \neq 98.6$
$\alpha = .05$
C.R. z < $-z_{.025}$ = -1.960
    z > $z_{.025}$ = 1.960

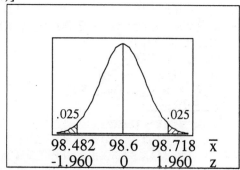

calculations:
$z_{\bar{x}} = (\bar{x} - \mu)/\sigma_{\bar{x}}$ $z_{\bar{x}} = (\bar{x} - \mu)/\sigma_{\bar{x}}$
$-1.960 = (\bar{x} - 98.6)/(.62/\sqrt{106})$ $1.960 = (\bar{x} - 98.6)/(.62/\sqrt{106})$
$\bar{x} = 98.482$ $\bar{x} = 98.718$
NOTE: There is no conclusion because this exercise is not asking for the completion of a

test of hypotheses.

a. $\beta = P(98.482 < \bar{x} < 98.718 \,|\, \mu = 98.7)$
   $= P(-3.62 < z < .30)$
   $= .4999 + .1179$
   $= .6178$

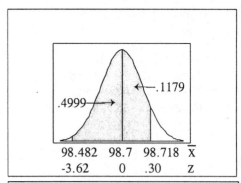

b. $\beta = P(98.482 < \bar{x} < 98.718 \,|\, \mu = 98.4)$
   $= P(1.36 < z < 5.28)$
   $= .4999 - .4131$
   $= .0868$

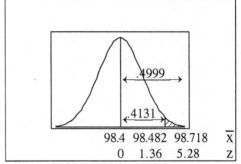

7-4 Testing a Claim about a Mean: Small Samples

1. original claim: $\mu = 75$ [$n \leq 30$ and σ unknown, use t]
   $H_o: \mu = 75$
   $H_1: \mu \neq 75$
   $\alpha = .05$
   C.R. $t < -t_{15,.025} = -2.132$
       $t > t_{15,.025} = 2.132$
   calculations:
     $t_{\bar{x}} = (\bar{x} - \mu)/s_{\bar{x}}$
       $= (77 - 75)/(15/\sqrt{16})$
       $= 2/3.75 = .533$

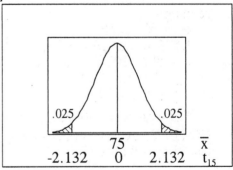

conclusion:
   Do not reject H_o; there is not sufficient evidence to reject the claim that $\mu = 75$.

3. original claim: $\mu > 2.00$ [$n \leq 30$ and σ unknown, use t]
   $H_o: \mu \leq 2.00$
   $H_1: \mu > 2.00$
   $\alpha = .01$
   C.R. $t > t_{23,.01} = 2.500$
   calculations:
     $t_{\bar{x}} = (\bar{x} - \mu)/s_{\bar{x}}$
       $= (2.35 - 2.00)/(.70/\sqrt{24})$
       $= .35/.1428 = 2.449$

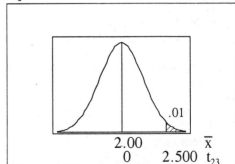

conclusion:
   Do not reject H_o; there is not sufficient evidence to conclude that $\mu > 2.00$.

5. since the calculated t = .533 is to the left of $t_{15,.25}$ = .691,
   the P-value for this two-tailed test is greater than 2·(.25)
   -- i.e., P-value > .50

7. since the calculated t = 2.449 is between $t_{23,.01}$ = 2.500 and $t_{23.025}$ = 2.069,
   the P-value for this one-tailed test is between .01 and .025
   -- i.e., .01 < P-value < .025

9. original claim: μ = 98.6 [n ≤ 30 and σ unknown, use t]
   H_o: μ = 98.6
   H_1: μ ≠ 98.6
   α = .05
   C.R. t < $-t_{24,.025}$ = -2.064
   $\quad$ t > $t_{24,.025}$ = 2.064
   calculations:
   $\quad t_{\bar{x}}$ = $(\bar{x} - \mu)/s_{\bar{x}}$
   $\qquad$ = (98.24 - 98.6)/(.56/√25)
   $\qquad$ = -.36/.112 = -3.214

   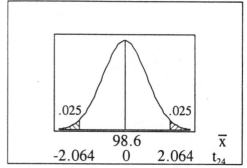

   conclusion:
   $\quad$ Reject H_o; there is sufficient evidence to reject the claim that μ = 98.6 and to conclude
   that μ ≠ 98.6 (in fact, that μ < 98.6).

11. original claim: μ < 1000 [n ≤ 30 and σ unknown, use t]
    summary statistics: n = 5, Σx = 3837, Σx^2 = 3268799, $\bar{x}$ = 767.4, s = 284.7
    H_o: μ ≥ 1000
    H_1: μ < 1000
    α = .05
    C.R. t < $-t_{4,.05}$ = -2.132
    calculations:
    $\quad t_{\bar{x}}$ = $(\bar{x} - \mu)/s_{\bar{x}}$
    $\qquad$ = (767.4 - 1000)/(284.7/√5)
    $\qquad$ = -232.6/127.3 = -1.827

    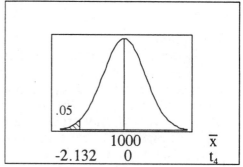

    conclusion:
    $\quad$ Do not reject H_o; there is not sufficient evidence to conclude that μ < 1000.
    No; since results like those obtained are not considered unusual if μ ≥ 1000, the company
    can not advertise with confidence that μ < 1000.

13. original claim: μ ≠ 243.5 [n ≤ 30 and σ unknown, use t]
    H_o: μ = 243.5
    H_1: μ ≠ 243.5
    α = .05
    C.R. t < $-t_{21,.025}$ = -2.080
    $\quad$ t > $t_{21,.025}$ = 2.080
    calculations:
    $\quad t_{\bar{x}}$ = $(\bar{x} - \mu)/s_{\bar{x}}$
    $\qquad$ = (170.2 - 243.5)/(35.3/√22)
    $\qquad$ = -73.3/7.526 = -9.740

    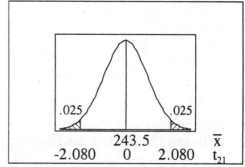

    conclusion:
    $\quad$ Reject H_o; there is sufficient evidence to conclude that μ ≠ 243.5 (in fact, that
    μ < 243.5).
    Yes; based on this result the VDT test scores are significantly lower than those for the
    printed version -- the VDT test is not fair to those choosing that format and should be
    revised.

15. original claim: $\mu \neq 41.9$ [$n \leq 30$ and σ unknown, use t]
    H_o: $\mu = 41.9$
    H_1: $\mu \neq 41.9$
    $\alpha = .01$
    C.R. $t < -t_{14,.005} = -2.977$
    $\quad\quad t > t_{14,.005} = 2.977$
    calculations:

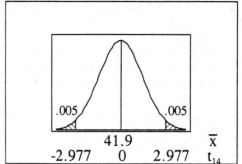

$$t_{\bar{x}} = (\bar{x} - \mu)/s_{\bar{x}}$$
$$= (31.0 - 41.9)/(10.5/\sqrt{15})$$
$$= -10.9/2.711 = -4.021$$

conclusion:
    Reject H_o; there is sufficient evidence to conclude that $\mu \neq 41.9$ (in fact, that $\mu < 41.9$).

17. original claim: $\mu > .9085$ [$n > 30$, use z (with s for σ)]
    H_o: $\mu \leq .9085$
    H_1: $\mu > .9085$
    $\alpha = .05$
    C.R. $z > z_{.05} = 1.645$
    calculations:

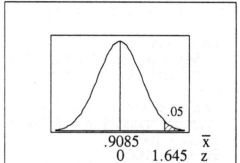

$$z_{\bar{x}} = (\bar{x} - \mu)/\sigma_{\bar{x}}$$
$$= (.9128 - .9085)/(.0395/\sqrt{33})$$
$$= .0043/.00688 = .625$$

conclusion:
    Do not reject H_o; there is not sufficient evidence to conclude that $\mu > .9085$.
    No; based on this sample we cannot conclude that the packages contain more than the claimed weight printed on the label. Even if the conclusion had been that $\mu > .9085$, that would not be evidence that the packages contain more than the printed weight because (1) the test involved only brown M&M's and (2) the assumption that packages contain 1498 M&M's is probably not correct -- it is more likely that the bags are filled by weight and not by count.

19. original claim: $\mu > 62.2$ [$n \leq 30$ and σ unknown, use t]
    summary statistics: $n = 9$, $\Sigma x = 579.4$, $\Sigma x^2 = 37334.60$, $\bar{x} = 64.38$, $s = 2.065$
    H_o: $\mu \leq 62.2$
    H_1: $\mu > 62.2$
    $\alpha = .01$
    C.R. $t > t_{8,.01} = 2.896$
    calculations:

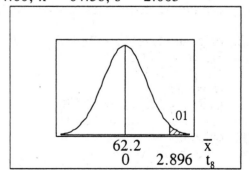

$$t_{\bar{x}} = (\bar{x} - \mu)/s_{\bar{x}}$$
$$= (64.38 - 62.2)/(2.065/\sqrt{9})$$
$$= 2.178/.6884 = 3.164$$

conclusion:
    Reject H_o; there is sufficient evidence to conclude that $\mu > 62.2$.

21. original claim: $\mu > 12$ [n > 30, use z (with s for σ)]
    summary statistics: n = 36, $\Sigma x = 439.2$, $\Sigma x^2 = 5358.36$, $\bar{x} = 12.20$, s = .05855
    H_o: $\mu \leq 12$
    H_1: $\mu > 12$
    $\alpha = .05$ [assumed]
    C.R. $z > z_{.05} = 1.645$
    calculations:
    $$z_{\bar{x}} = (\bar{x} - \mu)/\sigma_{\bar{x}}$$
    $$= (12.20 - 12)/(.05855/\sqrt{36})$$
    $$= .200/.009759 = 20.494$$

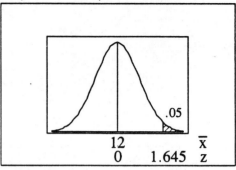

    conclusion:
    Reject H_o; there is sufficient evidence to conclude that $\mu > 12$.

23. original claim: $\mu = 3.39$ [n ≤ 30 and σ unknown, use t]
    summary statistics: n = 16, $\Sigma x = 58.80$, $\Sigma x^2 = 222.5710$, $\bar{x} = 3.675$, s = .657
    H_o: $\mu = 3.39$
    H_1: $\mu \neq 3.39$
    $\alpha = .05$
    C.R. $t < -t_{15,.025} = -2.132$
    $\quad\quad t > t_{15,.025} = 2.132$
    calculations:
    $$t_{\bar{x}} = (\bar{x} - \mu)/s_{\bar{x}}$$
    $$= (3.675 - 3.39)/(.657/\sqrt{16})$$
    $$= .285/.1643 = 1.734$$

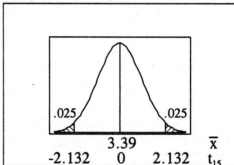

    conclusion:
    Do not reject H_o; there is not sufficient evidence to conclude that $\mu \neq 3.39$ -- i.e., there is not sufficient evidence to conclude that the vitamin supplement has an effect on birth weight.

25. original claim: $\mu < 1000$ [n ≤ 30 and σ unknown, use t]
    H_o: $\mu \geq 1000$
    H_1: $\mu < 1000$
    $\alpha = .05$ [assumed]
    C.R. $t < -t_{20,.05} = -1.725$
    calculations:
    $$t_{\bar{x}} = (\bar{x} - \mu)/s_{\bar{x}}$$
    $$= (985.67 - 1000)/(46.52/\sqrt{21})$$
    $$= -1.412 \text{ [from TI-83 Plus]}$$

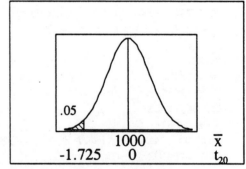

    conclusion:
    Do not reject H_o; there is not sufficient evidence to conclude that $\mu < 1000$.
    P-value = $P(t_{20} < -1.412) = .08666$ [from TI-83 Plus]

27. original claim: $\mu > 12$ [$n \le 30$ and σ unknown, use t]
H_o: $\mu \le 12$
H_1: $\mu > 12$
$\alpha = .01$
C.R. $t > t_{10,.01} = 2.764$
calculations:
$t_{\bar{x}} = (\bar{x} - \mu)/s_{\bar{x}}$
$= (13.286 - 12)/(1.910/\sqrt{11})$
$= 1.286/.576 = 2.23$ [from Minitab]

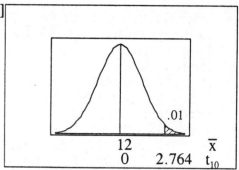

conclusion:
Do not reject H_o; there is not sufficient evidence to conclude that $\mu > 12$.
P-value $= .025$ [from Minitab]

29. Only the P-value is changed. The summary values n, $\bar{x}$ and s do not change. The standard error of the mean, which is $s_{\bar{x}} = s/\sqrt{n}$, does not change. The calculated t, which is $t_{\bar{x}} = (\bar{x} - \mu)/s_{\bar{x}}$, does not change. Since the test is now two-tailed, the P-value doubles and is now $2 \cdot P(t_{14} > 3.17) = 2 \cdot (.0034) = .0068$.

7-5 Testing a Claim about a Proportion

NOTE: To be consistent with the notation of the previous sections, and thereby reinforcing the patterns and concepts presented in those sections, the manual uses the "usual" z formula written to apply to $\hat{p}$'s
$z_{\hat{p}} = (\hat{p} - \mu_{\hat{p}})/\sigma_{\hat{p}}$
When the normal approximation to the binomial applies, the $\hat{p}$'s are normally distributed
with $\mu_{\hat{p}} = p$ and $\sigma_{\hat{p}} = \sqrt{pq/n}$
And so the formula for the z statistic may also be written as
$z_{\hat{p}} = (\hat{p} - p)/\sqrt{pq/n}$

1. original claim: $p < .20$ [normal approximation to the binomial, use z]
$\hat{p} = x/n = 149/880 = .169$
H_o: $p \ge .20$
H_1: $p < .20$
$\alpha = .05$
C.R. $z < -z_{.05} = -1.645$
calculations:
$z_{\hat{p}} = (\hat{p} - \mu_{\hat{p}})/\sigma_{\hat{p}}$
$= (.169 - .20)/\sqrt{(.20)(.80)/880}$
$= -.0307/.0135 = -2.275$

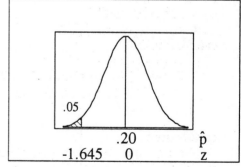

conclusion:
Reject H_o; there is sufficient evidence to conclude that $p < .20$.
P-value $= P(z < -2.285) = .5000 - .4887 = .0113$.

No; since technology and the use of technology is changing so rapidly, any figures from 1997 would no longer be valid today. [By the time you are reading this, e-mail could be so common that essentially 100% of the households use it -- or it could have been replaced by something newer and be so out-dated that essentially 0% of the households still use it!]

3. original claim: p > .50 [normal approximation to the binomial, use z]
   $\hat{p} = x/n = 544/850 = .640$
   $H_o: p \le .50$
   $H_1: p > .50$
   $\alpha = .02$
   C.R. $z > z_{.02} = 2.05$
   calculations:
   $z_{\hat{p}} = (\hat{p} - \mu_{\hat{p}})/\sigma_{\hat{p}}$
   $= (.640 - .50)/\sqrt{(.50)(.50)/850}$
   $= .140/.0171 = 8.163$

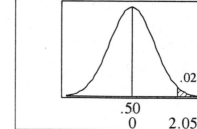

conclusion:
   Reject H_o; there is sufficient evidence to conclude that p > .50.
P-value = P(z > 8.16) = .5000 - .4999 = .0001.
This test provides evidence that most drivers change tapes or CD's while driving, but it does not address the question as to whether that act is a driving hazard. The actual percent of drivers engaging in some practice is not necessarily what determines whether it is a hazard. If 2% of the drivers are intoxicated, that would be a hazard; if 98% of the drivers are chewing gum, that would not be a hazard.

5. original claim: p = .43 [normal approximation to the binomial, use z]
   $\hat{p} = x/n = 308/611 = .504$
   $H_o: p = .43$
   $H_1: p \ne .43$
   $\alpha = .04$
   C.R. $z < -z_{.02} = -2.05$
   $z > z_{.02} = 2.05$
   calculations:
   $z_{\hat{p}} = (\hat{p} - \mu_{\hat{p}})/\sigma_{\hat{p}}$
   $= (.504 - .43)/\sqrt{(.43)(.57)/611}$
   $= .0741/.0200 = 3.699$

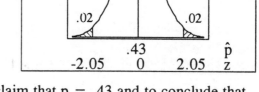

conclusion:
   Reject H_o; there is sufficient evidence reject the claim that p = .43 and to conclude that p ≠ .43 (in fact, that p > .43).
P-value = 2·P(z > 3.70) = 2·(.5000 - .4999) = 2·(.0001) = .0002
The results suggest that people are not telling the truth. Most likely (1) some people who voted for the loser don't want to reveal their true vote and be identified with a loser and (2) some people who didn't vote at all won't admit that fact and claim to have voted for the winner.

7. original claim: p = .01 [normal approximation to the binomial, use z]
   $\hat{p} = x/n = 20/1234 = .0162$
   $H_o: p = .01$
   $H_1: p \ne .01$
   $\alpha = .05$
   C.R. $z < -z_{.025} = -1.960$
   $z > z_{.025} = 1.960$
   calculations:
   $z_{\hat{p}} = (\hat{p} - \mu_{\hat{p}})/\sigma_{\hat{p}}$
   $= (.0162 - .01)/\sqrt{(.01)(.99)/1234}$
   $= .0062/.0028 = 2.192$

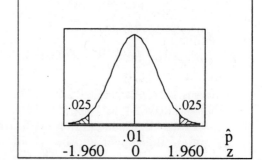

conclusion:
   Reject H_o; there is sufficient evidence reject the claim that p = .01 and to conclude that p ≠ .01 (in fact, that p > .01).
P-value = 2·P(z > 2.19) = 2·(.5000 - .4857) = 2·(.0143) = .0286
No; based on these results, scanners do not appear to help consumers avoid overcharges.

9. original claim: p > .50 [normal approximation to the binomial, use z]
   NOTE: A "success" is <u>not</u> being knowledgeable about the Holocaust.
   $\hat{p} = x/n = 268/506 = .530$
   $H_o: p \leq .50$
   $H_1: p > .50$
   $\alpha = .05$
   C.R. $z > z_{.05} = 1.645$
   calculations

   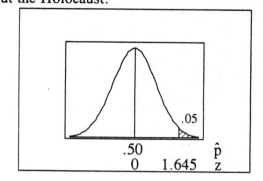

   $$z_{\hat{p}} = (\hat{p} - \mu_{\hat{p}})/\sigma_{\hat{p}}$$
   $$= (.530 - .50)/\sqrt{(.50)(.50)/506}$$
   $$= .0296/.0222 = 1.334$$

   conclusion:
   Do not reject H_o; there is not sufficient evidence to conclude that p > .50.
   P-value = P(z > 1.33) = .5000 - .4082 = .0918
   No; we cannot be 95% certain that p > .50 and that the curriculum needs to be revised.

11. original claim: p > .50 [normal approximation to the binomial, use z]
    NOTE: A "success" is <u>not</u> having quit smoking one year later.
    $\hat{p} = x/n = 39/71 = .5493$
    $H_o: p \leq .50$
    $H_1: p > .50$
    $\alpha = .10$
    C.R. $z > z_{.10} = 1.282$
    calculations

    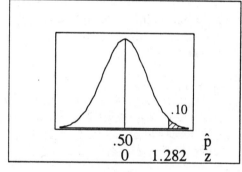

    $$z_{\hat{p}} = (\hat{p} - \mu_{\hat{p}})/\sigma_{\hat{p}}$$
    $$= (.5493 - .50)/\sqrt{(.50)(.50)/71}$$
    $$= .0493/.0593 = .831$$

    conclusion:
    Do not reject H_o; there is not sufficient evidence to conclude that p > .50.
    P-value = P(z > .83) = .5000 - .2967 = .2033
    No; while the sample success rate is less than ½, that value must be seen in context --
    would you call a major league baseball player with a .475 batting average a failure because
    his success rate is less than ½? The success rate must be compared to success rates of
    other programs (or to a control group) and to the fact that helping even a few is of some
    effect.

13. original claim: p = .75 [normal approximation to the binomial, use z]
    $\hat{p} = x/n = 410/500 = .82$ NOTE: x = (.82)(500) = 410
    $H_o: p = .75$
    $H_1: p \neq .75$
    $\alpha = .05$ [assumed]
    C.R. $z < -z_{.025} = -1.960$
         $z > z_{.025} = 1.960$
    calculations:

    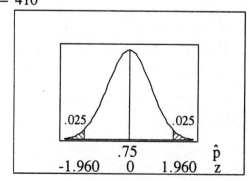

    $$z_{\hat{p}} = (\hat{p} - \mu_{\hat{p}})/\sigma_{\hat{p}}$$
    $$= (.82 - .75)/\sqrt{(.75)(.25)/500}$$
    $$= .07/.0194 = 3.615$$

    conclusion:
    Reject H_o; there is sufficient evidence reject the claim that p = .75 and to conclude that
    p ≠ .75 (in fact, that p > .75).
    P-value = 2·P(z > 3.61) = 2·(.5000 - .4999) = 2·(.0001) = .0002

15. original claim: p < .10 [normal approximation to the binomial, use z]
    $\hat{p} = x/n = 28/400 = .07$ NOTE: x = (.07)(400) = 28
    H_o: p ≥ .10
    H_1: p < .10
    α = .01
    C.R. z < $-z_{.01}$ = -2.327
    calculations:
    $z_{\hat{p}} = (\hat{p} - \mu_{\hat{p}})/\sigma_{\hat{p}}$
    $= (.07 - .10)/\sqrt{(.10)(.90)/400}$
    $= -.03/.0150 = -2.000$

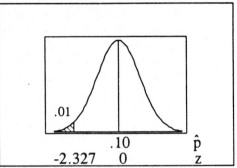

conclusion:
    Do not reject H_o; there is not sufficient evidence to conclude that p < .10.
    P-value = P(z < -2.00) = .5000 - .4772 = .0228
The conclusion cannot be applied to all U.S. college students. The students surveyed were from one particular college and were not a random sample of all U.S. college students.

17. original claim: p > .75 [normal approximation to the binomial, use z]
    $\hat{p} = x/n = 455/500 = .91$ NOTE: x = (.91)(500) = 455
    H_o: p ≤ .75
    H_1: p > .75
    α = .01
    C.R. z > $z_{.01}$ = 2.327
    calculations:
    $z_{\hat{p}} = (\hat{p} - \mu_{\hat{p}})/\sigma_{\hat{p}}$
    $= (.91 - .75)/\sqrt{(.75)(.25)/500}$
    $= .16/.0194 = 8.262$

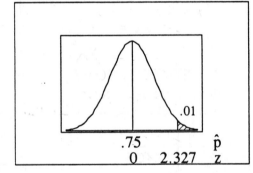

conclusion:
    Reject H_o; there is sufficient evidence to conclude that p > .75.
    P-value = P(z > 8.26) = .5000 - .4999 = .0001
Yes; based on the sample results, the funding will be approved.

19. original claim: p = .10 [normal approximation to the binomial, use z]
    $\hat{p} = x/n = 5/100 = .05$
    H_o: p = .10
    H_1: p ≠ .10
    α = .05 [assumed]
    C.R. z < $-z_{.025}$ = -1.960
        z > $z_{.025}$ = 1.960
    calculations:
    $z_{\hat{p}} = (\hat{p} - \mu_{\hat{p}})/\sigma_{\hat{p}}$
    $= (.05 - .10)/\sqrt{(.10)(.90)/100}$
    $= -.05/.030 = -1.667$

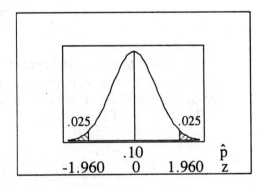

conclusion:
    Do no reject H_o; there is not sufficient evidence reject the claim that p = .10
    P-value = 2·P(z < -1.67) = 2·(.5000 - .4525) = 2·(.0475) = .0950

21. a. For the claim $p > .0025$, we have $H_0 : p \le .0025$.
    Since $np = (80)(.0025) = 0.2 < 5$, the normal approximation to the binomial is not appropriate and the methods of this section cannot be used.

   b. binomial: $n = 80$, $p = .0025$
      $P(x) = [n!/x!(n-x)!]p^x(1-p)^{n-x}$
      $P(x = 0) = [80!/0!80!](.0025)^0(.9975)^{80} = .818525754$
      $P(x = 1) = [80!/1!79!](.0025)^1(.9975)^{79} = .164115439$
      $P(x = 2) = [80!/2!78!](.0025)^2(.9975)^{78} = .016247017$
      $P(x = 3) = [80!/3!77!](.0025)^3(.9975)^{77} = .001058703$
      $P(x = 4) = [80!/4!76!](.0025)^4(.9975)^{76} = .000051078$
      $P(x = 5) = [80!/5!75!](.0025)^5(.9975)^{75} = .000001946$
      $P(x = 6) = [80!/6!74!](.0025)^6(.9975)^{74} = \underline{.000000061}$
      $\phantom{P(x = 6) = [80!/6!74!](.0025)^6(.9975)^{74} = } .999999998$

      $P(x \ge 7) = 1 - P(x \le 6)$
      $ = 1 - .999999998$
      $ = .000000002$

   c. Based on the result from part (b), getting 7 or more males with this color blindness if $p = .0025$ would be an extremely rare event. This leads to reject the claim that $p = .0025$.
      NOTE: In a one-tailed test of $H_0 : p = .0025$ vs. $H_1 : p > .0025$, the observed value $x = 7$ produces:
      P-value $= P(\text{getting a result as extreme as or more extreme than } x = 7)$
      $\phantom{\text{P-value}} = P(x \ge 7)$
      $\phantom{\text{P-value}} = .000000002$

ADDITIONAL NOTE: This problem may be worked in a format similar to that used in this section as follows.
original claim: $p > .0025$ [$np = (80)(.0025) = 0.2 < 5$; use binomial distribution]

$x = 7$
$H_0 : p \le .0025$
$H_1 : p > .0025$
$\alpha = .01$
C.R. $x > 2$
calculations
$\quad x = 7$

NOTE: Refer to this probability distribution
for $n=80$ and $p=.0025$

x	P(x)
0	.818526
1	.164115
2	.016247
3	.001059
4	.000051
5	.000002
6	0$^+$
7	0$^+$
⋮	⋮
	$\overline{1.000000}$

conclusion:
   Reject H_0; there is sufficient evidence to conclude $p > .0025$.
RATIONALE: We want to place .01 (or as close to it as possible without going over) in the upper tail of this one-tailed test.
   $P(x > 1) = P(x = 2) + \ldots$
   $ = .016247 + \ldots$
   $ > .01$
   $P(x > 2) = 1 - [P(x = 0) + P(x = 1) + P(x = 2)]$
   $ = 1 - [.818526 + .164115 + .016247]$
   $ = 1 - .998888$
   $ = .001112 \le .01$; hence this is the desired critical region

7-6 Testing a Claim about a Standard Deviation or Variance

NOTE: Following the pattern used with the z and t distributions, this manual uses the closest entry from Table A-4 for χ^2 as if it were the precise value necessary and does not use interpolation. This procedure sacrifices very little accuracy -- and even interpolation does not yield precise values. When extreme accuracy is needed in practice, statisticians refer either to more accurate tables or to computer-produced values.

ADDITIONAL NOTE: The χ^2 distribution depends upon n, and it "bunches up" around df = n-1. In addition, the formula $\chi^2 = (n-1)s^2/\sigma^2$ used in the calculations contains df = n-1. When the exact df needed in the problem does not appear in the table and the closest χ^2 value is used to determine the critical region, some instructors recommend using the same df in the calculations that were used to determine the C.R. This manual typically uses the closest entry to determine the C.R. and the n from the problem in the calculations, even though this introduces a slight discrepancy.

1. a. $\chi^2_{9,.995} = 1.735$ b. $\chi^2_{26,.01} = 45.642$ c. $\chi^2_{20,.95} = 10.851$
   $\chi^2_{9,.005} = 23.589$

GRAPHICS NOTE: To illustrate χ^2 tests of hypotheses, this manual uses a "generic" figure, resembling a chi-squared distribution with approximately 4 degrees of freedom – chi-squared distributions with 1 and 2 degrees of freedom actually have no upper limit and approach the y axis asymptotically, while chi-squared distributions with more than 30 degrees of freedom are essentially symmetric and normal-looking. The expected value of the chi-squared distribution is the degrees of freedom for the problem, typically n-1. Since the distribution is positively skewed, this manual indicates that value slightly to the right of the figure's "peak." Loosely speaking, the distribution "bunches up" around the degrees of freedom.

3. original claim: $\sigma = 15$
   H_o: $\sigma = 15$
   H_1: $\sigma \neq 15$
   $\alpha = .05$
   C.R. $\chi^2 < \chi^2_{9,.975} = 2.700$
   $\chi^2 > \chi^2_{9,.025} = 19.023$
   calculations:
   $\chi^2 = (n-1)s^2/\sigma^2$
   $= (9)(18.0)^2/(15)^2 = 12.960$

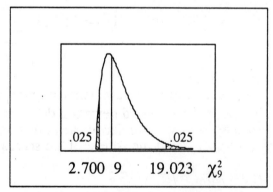

conclusion:
   Do not reject H_o; there is not sufficient evidence to reject the claim that $\sigma = 15$.

5. original claim: $\sigma > .04$
   H_o: $\sigma \leq .04$
   H_1: $\sigma > .04$
   $\alpha = .01$
   C.R. $\chi^2 > \chi^2_{39,.01} = 63.691$
   calculations:
   $\chi^2 = (n-1)s^2/\sigma^2$
   $= (39)(.31)^2/(.04)^2 = 2342.4375$

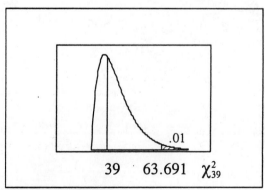

conclusion:
   Reject H_o; there is sufficient evidence to conclude that $\sigma > .04$.
We expect more variation for the weights of peanut M&M's. Plain M&M's are produced

according to specifications and should show little variation from piece to piece. Peanut M&M's start with a peanut -- and peanuts are not produced to specifications, but occur with varying sizes. Even though the company rejects very large and very small peanuts, there will still be more variation from piece to piece than there would be from a controlled production process.

7. original claim: $\sigma \neq 43.7$
   H_o: $\sigma = 43.7$
   H_1: $\sigma \neq 43.7$
   $\alpha = .05$
   C.R. $\chi^2 < \chi^2_{80,.975} = 57.153$
   $\quad\quad \chi^2 > \chi^2_{80,.025} = 106.629$
   calculations:
   $\quad \chi^2 = (n-1)s^2/\sigma^2$
   $\quad\quad = (80)(52.3)^2/(43.7)^2 = 114.586$

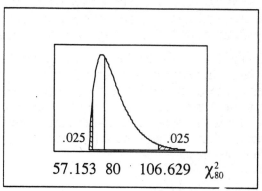

   conclusion:
       Reject H_o; there is sufficient evidence to conclude that $\sigma \neq 43.7$ (in fact, that $\sigma > 43.7$).

   Since a larger σ indicates more variability, it appears that the new production method is worse.

9. original claim: $\sigma < 14.1$
   H_o: $\sigma \geq 14.1$
   H_1: $\sigma < 14.1$
   $\alpha = .01$
   C.R. $\chi^2 < \chi^2_{26,.99} = 12.198$
   calculations:
   $\quad \chi^2 = (n-1)s^2/\sigma^2$
   $\quad\quad = (26)(9.3)^2/(14.1)^2 = 11.311$

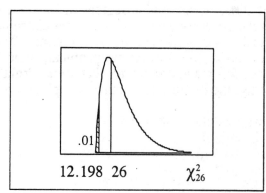

   conclusion:
       Reject H_o; there is sufficient evidence to conclude that $\sigma < 14.1$.

   No; the lower standard deviation does not suggest that the current class is doing better, but that they are more similar to each other and show less variation. The standard deviation speaks only about the spread of the scores and not their location.

11. original claim: $\sigma > 19.7$
    H_o: $\sigma \leq 19.7$
    H_1: $\sigma > 19.7$
    $\alpha = .05$
    C.R. $\chi^2 > \chi^2_{49,.05} = 67.505$
    calculations:
    $\quad \chi^2 = (n-1)s^2/\sigma^2$
    $\quad\quad = (49)(23.4)^2/(19.7)^2 = 69.135$

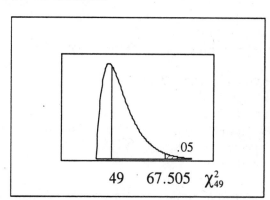

    conclusion:
        Reject H_o; there is sufficient evidence to conclude that $\sigma > 19.7$.

13. summary statistics: $n = 25$, $\Sigma x = 1721.07$, $\Sigma x^2 = 118614.7057$, $\bar{x} = 68.843$, $s^2 = 5.476$
    original claim: $\sigma \neq 2.9$
    H_o: $\sigma = 2.9$
    H_1: $\sigma \neq 2.9$
    $\alpha = .05$ [assumed]
    C.R. $\chi^2 < \chi^2_{24,.975} = 12.401$
    $\qquad \chi^2 > \chi^2_{24,.025} = 39.364$
    calculations:
    $\quad \chi^2 = (n-1)s^2/\sigma^2$
    $\qquad = (24)(5.476)/(2.9)^2 = 15.627$

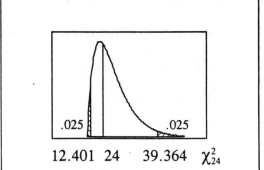

conclusion:
    Do not reject H_o; there is not sufficient evidence to conclude that $\sigma \neq 2.9$.

15. summary statistics: $n = 9$, $\Sigma x = 1089$, $\Sigma x^2 = 132223$, $\bar{x} = 121.0$, $s^2 = 56.75$
    original claim: $\sigma < 29$
    H_o: $\sigma \geq 29$
    H_1: $\sigma < 29$
    $\alpha = .01$
    C.R. $\chi^2 < \chi^2_{8,.99} = 1.646$
    calculations:
    $\quad \chi^2 = (n-1)s^2/\sigma^2$
    $\qquad = (8)(56.75)/(29)^2 = .540$

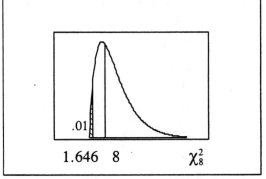

conclusion:
    Reject H_o; there is sufficient evidence to conclude that $\sigma < 29$.

17. a. To find the P-value for $\chi^2_9 = 12.960$ in a two-tailed test,
    note that $\chi^2_{9,.10} = 14.684$ is the smallest value above the "center" at $\chi^2_9 = 9.000$
    Therefore, $\frac{1}{2}$(P-value) $> .10$ and P-value $> .20$.

    b. To find the P-value for $\chi^2_{39} = 2342.4375$ in a one-tailed greater than test,
    note that $\chi^2_{40,.005} = 66.766$ is the largest χ^2 in the row.
    Therefore, P-value $< .005$.

    c. To find the P-value for $\chi^2_8 = .540$ in a one-tailed less than test,
    note that $\chi^2_{8,.995} = 1.344$ is the smallest χ^2 in the row.
    Therefore, P-value $< .005$

Review Exercises

1. a. Prefer a P-value of .001. This indicates only a .1% chance of achieving the observed
    effectiveness by chance alone if the new cure really has no effect.
    b. The original claim should be $\mu < 114.8$. This would lead to
    H_o: $\mu \geq 114.8$
    H_1: $\mu < 114.8$
    c. The original claim that $\mu < 90$ leads to
    H_o: $\mu \geq 90$
    H_1: $\mu < 90$
    d. Fail to reject H_o: $\mu \leq 12$; there is not enough evidence to conclude the original claim
    that $\mu > 12$.

e. A type I error is the mistake of rejecting the null hypothesis when the null hypothesis is really true.

2. a. concerns μ: $n \leq 30$ and σ unknown, use t
$-t_{9,.05} = -1.833$
   b. concerns p: normal approx. to the binomial, use z
$\pm z_{.005} = \pm 2.575$
   c. concerns σ: use χ^2
$\chi^2_{19,.975} = 8.907$ and $\chi^2_{19,.025} = 32.852$
   d. concerns μ: $n > 30$, use z (with s for σ)
$\pm z_{.05} = \pm 1.645$
   e. concerns p: normal approx. to the binomial, use z
$-z_{.01} = -2.327$

3. original claim: $p > .50$ [concerns p: normal approximation to the binomial, use z]
$\hat{p} = x/n = x/504 = .57$ NOTE: $x=(.57)(504)=287.28$; any $285 \leq x \leq 289$ rounds to 57%
H_o: $p \leq .50$
H_1: $p > .50$
$\alpha = .01$
C.R. $z > z_{.01} = 2.327$
calculations:
$z_{\hat{p}} = (\hat{p} - \mu_p)/\sigma_{\hat{p}}$
$= (.57 - .50)/\sqrt{(.50)(.50)/504}$
$= .07/.0223 = 3.143$

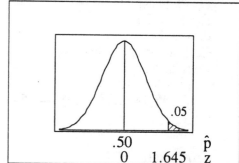

conclusion:
Reject H_o; there is sufficient evidence to conclude that $p > .50$.
P-value = $P(z > 3.14) = .5000 - .4999 = .0001$

4. original claim: $\mu < 12$ [concerns μ: $n \leq 30$ and σ unknown, use t]
H_o: $\mu \geq 12$
H_1: $\mu < 12$
$\alpha = .05$ [assumed]
C.R. $t < -t_{23,.05} = -1.714$
calculations:
$t_{\bar{x}} = (\bar{x} - \mu)/s_{\bar{x}}$
$= (11.4 - 12)/(.62/\sqrt{24})$
$= -.6/.1266 = -4.741$

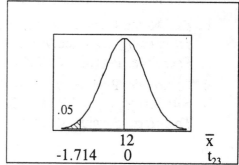

conclusion:
Reject H_o; there is sufficient evidence to conclude that $\mu < 12$.
No; assuming that proper random sampling techniques were used, his claim that the sample is too small to be meaningful is not valid. The t distribution adjusts for the various sample sizes to produce a valid test for any sample size n.

5. original claim: $\sigma < .75$ [concerns σ: use χ^2]
    H_o: $\sigma \geq .75$
    H_1: $\sigma < .75$
    $\alpha = .05$
    C.R. $\chi^2 < \chi^2_{60,.95} = 43.188$
    calculations:
    $$\chi^2 = (n-1)s^2/\sigma^2$$
    $$= (60)(.48)^2/(.75)^2 = 24.576$$

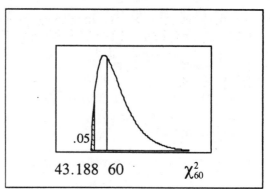

    conclusion:
        Reject H_o; there is sufficient evidence to conclude that $\sigma < 29$.
    The weights appear to be more consistent (i.e., they have a smaller standard deviation and less variation from bag to bag) now that the worn parts have been replaced.

6. original claim: $p < .10$ [concerns p: normal approximation to the binomial, use z]
    NOTE: $x=(.08)(1248)=99.84$; any $94 \leq x \leq 106$ rounds to 8%
    $\hat{p} = x/n = x/1248 = .08$
    H_o: $p \geq .10$
    H_1: $p < .10$
    $\alpha = .01$
    C.R. $z < -z_{.01} = -2.327$
    calculations:
    $$z_{\hat{p}} = (\hat{p} - \mu_{\hat{p}})/\sigma_{\hat{p}}$$
    $$= (.08 - .10)/\sqrt{(.10)(.90)/1248}$$
    $$= -.02/.00849 = -2.355$$

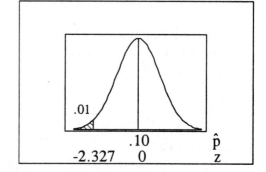

    conclusion:
        Reject H_o; there is sufficient evidence to conclude that $p < .10$.
    Yes; based on this result, the phrase "almost 1 out of 10" is justified. While "almost" is not well-defined, 10% is the closest round number to 8%. The test indicates we are confident that $p < .10$, but does not specify how much less.

    ADDITIONAL NOTE: Using the upper possibility $\hat{p} = x/n = 106/1248 = .0849$ from the previous NOTE produces $z_{\hat{p}} = -1.774$, which is not enough evidence to reject H_o: $p \geq .10$. This not only supports the opinion that "almost 1 out of 10" is justified, but it also shows how sensitive tests may be to rounding procedures -- and that in this case the results are actually not reported with sufficient accuracy to conduct the test.

7. original claim: $\mu < 7124$ [concerns μ: $n > 30$, use z (with s for σ)]
    H_o: $\mu \geq 7124$
    H_1: $\mu < 7124$
    $\alpha = .01$
    C.R. $z < -z_{.01} = -2.327$
    calculations:
    $$z_{\bar{x}} = (\bar{x} - \mu)/\sigma_{\bar{x}}$$
    $$= (6047 - 7124)/(2944/\sqrt{750})$$
    $$= -1077/107.50 = -10.019$$

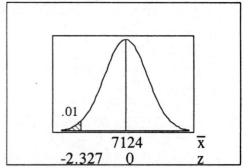

    conclusion:
        Reject H_o; there is sufficient evidence to conclude that $\mu < 7124$.
    Not necessarily; distance driven is only factor to take into consideration when determining insurance rates. Even though the 16-24 year olds drive less than those in the higher age bracket, their lack of experience and maturity may still make them higher insurance risks.

8. original claim: $p \geq .50$ [concerns p: normal approximation to the binomial, use z]
   NOTE: $x = (.28)(1012) = 283.36$; any $279 \leq x \leq 288$ rounds to 28%
   $\hat{p} = x/n = x/1012 = .28$
   H_o: $p \geq .50$
   H_1: $p < .50$
   $\alpha = .01$
   C.R. $z < -z_{.01} = -2.327$
   calculations:
   $z_{\hat{p}} = (\hat{p} - \mu_{\hat{p}})/\sigma_{\hat{p}}$
   $= (.28 - .50)/\sqrt{(.50)(.50)/1012}$
   $= -.22/.0157 = -13.997$

   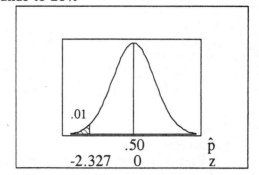

   conclusion:
   Reject H_o; there is sufficient evidence to reject the claim that $p \geq .50$ and to conclude that $p < .50$.

   Yes; based on this result, it appears that fruitcake producers should be concerned. If a product is not perceived as useful or appropriate for its intended purpose, its sales will probably fall if the economy weakens and/or new alternatives become available.

9. summary statistics: $n = 20$, $\Sigma x = 465.7$, $\Sigma x^2 = 11234.37$, $\overline{x} = 23.285$, $s = 4.534$
   original claim: $\mu = 20.0$ [concerns μ: $n \leq 30$ and σ unknown, use t]
   H_o: $\mu = 20.0$
   H_1: $\mu \neq 20.0$
   $\alpha = .01$
   C.R. $t < -t_{19,.005} = -2.861$
   $t > t_{19,.005} = 2.861$
   calculations:
   $t_{\overline{x}} = (\overline{x} - \mu)/s_{\overline{x}}$
   $= (23.285 - 20.0)/(4.534/\sqrt{20})$
   $= 3.285/1.014 = 3.240$

   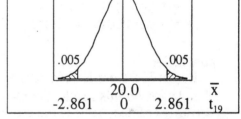

   conclusion:
   Reject H_o; there is sufficient evidence to reject the claim that $\mu = 20.0$ and to conclude that $\mu \neq 20.0$ (in fact, that $\mu > 20.0$).

   No; these pills are not acceptable at the .01 level because their mean amount appears to be higher than the stated 20.0 mg. Even if the hypothesis that $\mu = 20.0$ had not been rejected, the pills may be unacceptable because of their variability. While no specific criteria for variability were stated, note that the pills range from 165.0% (33 mg.) to 79.9% (15.9 mg.) of their stated amount.

10. original claim: $\mu > 30,000$ [concerns μ: $n > 30$, use z (with s for σ)]
    H_o: $\mu \leq 30,000$
    H_1: $\mu > 30,000$
    $\alpha = .01$
    C.R. $z > z_{.01} = 2.327$
    calculations:
    $z_{\overline{x}} = (\overline{x} - \mu)/\sigma_{\overline{x}}$
    $= (30122 - 30000)/(14276/\sqrt{150})$
    $= 122/1165.6 = .105$

    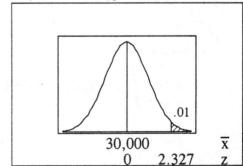

    conclusion:
    Do not reject H_o; there is not sufficient evidence to conclude that $\mu > 30,000$.

    No; her claim is not justified. Even though the sample results appear to support her claim, such results would not be unusual even if her claim were false -- and so the results are not enough to conclude with appropriate confidence that her claim is correct.

Cumulative Review Exercises

1. scores in order: 410 440 460 490 570 630 720 780
   summary statistics: $n = 8$, $\Sigma x = 4500$, $\Sigma x^2 = 2662000$
   a. $\bar{x} = (\Sigma x)/n = 4500/8 = 562.5$
   b. $\tilde{x} = (490 + 570)/2 = 530.0$
   c. $s^2 = [n(\Sigma x^2) - (\Sigma x)^2]/[n(n-1)]$
      $= [8(2662000) - (4500)^2]/[8(7)]$
      $= 18678.6$
      $s = 136.7$
   d. $s^2 = 186 / 8.6$
   e. $R = 780 - 410 = 370$
   f. $\bar{x} = \pm t_{7,.025}s_{\bar{x}}$
      $562.5 \pm 2.365 \cdot (136.7/\sqrt{8})$
      562.5 ± 114.3
      $448.2 < \mu < 676.8$
   g. original claim: $\mu = 496$ [concerns μ: $n \le 30$ and σ unknown, use t]
      $H_o: \mu = 496$
      $H_1: \mu \ne 496$
      $\alpha = .05$
      C.R. $t < -t_{7,.025} = -2.365$
          $t > t_{7,.025} = 2.365$
      calculations:
      $t_{\bar{x}} = (\bar{x} - \mu)/s_{\bar{x}}$
          $= (562.5 - 496)/(136.7/\sqrt{8})$
          $= 66.5/48.3 = 1.376$

   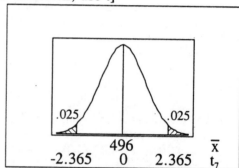

   conclusion:
      Do not reject H_o; there is not sufficient evidence to reject the claim that $\mu = 496$.
   h. Based on the preceding results, it appears that the SAT math scores of female applicants
      to the College of Newport are about the same as those of the general female population.
      Both the confidence interval in (f) and the test in (g) fail to reject $\mu = 496$ as a
      possibility. In addition, ½ of the sample falls below 496 and ½ of the sample falls
      above 496.

2. normal distribution: $\mu = 496$, $\sigma = 108$
   a. $P(x > 500)$
      $= P(z > .04)$
      $= .5000 - .0160$
      $= .4840$

   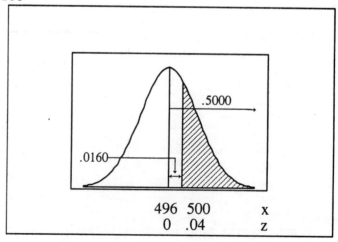

b. let A = a selected score is above 500

  P(A) = .4840, for each selection

  P(A_1 and A_2 and A_3 and A_4 and A_5)

  = P(A_1)·P(A_2)·P(A_3)·P(A_4)·P(A_4)

  = $(.4840)^5$

  = .0266

c. for samples of size n = 5,

  $\mu_{\bar{x}} = \mu = 496$

  $\sigma_{\bar{x}} = \sigma/\sqrt{n} = 108/\sqrt{5} = 48.30$

  normal distribution, since the original distribution is so

  P($\bar{x} > 500$)

  = P(z > .08)

  = .5000 - .0319

  = .4681

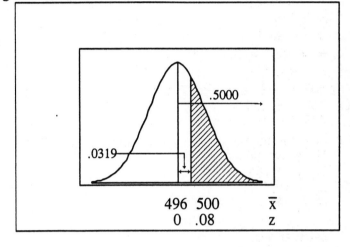

d. for P_{90}, A = .4000 and z = 1.282

  x = $\mu + z·\sigma$

  = 496 + (1.282)(108)

  = 496 + 138

  = 634

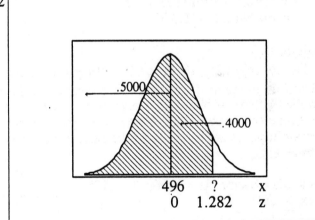

Chapter 8

Inferences from Two Samples

8-2 Inferences about Two Means: Independent and Large Samples

NOTE: To be consistent with the notation of the previous sections, and thereby reinforcing the patterns and concepts presented in those sections, the manual uses the "usual" z formula written to apply to $\bar{x}_1$-$\bar{x}_2$'s

$$z_{\bar{x}_1 - \bar{x}_2} = (\bar{x}_1 - \bar{x}_2 - \mu_{\bar{x}_1 - \bar{x}_2})/\sigma_{\bar{x}_1 - \bar{x}_2}$$

with $\mu_{\bar{x}_1 - \bar{x}_2} = \mu_1 - \mu_2$

and $\sigma_{\bar{x}_1 - \bar{x}_2} = \sqrt{\sigma_1^2/n_1 + \sigma_2^2/n_2}$

And so the formula for the z statistic may also be written as

$$z_{\bar{x}_1 - \bar{x}_2} = ((\bar{x}_1 - \bar{x}_2) - (\mu_1 - \mu_2))/\sqrt{\sigma_1^2/n_1 + \sigma_2^2/n_2}$$

1. original claim: $\mu_1 - \mu_2 = 0$ [$n_1 > 30$ and $n_2 > 30$, use z (with s's for σ's)]
    $\bar{x}_1 - \bar{x}_2 = 7.00 - 6.00 = 1.00$
    H_0: $\mu_1 - \mu_2 = 0$
    H_1: $\mu_1 - \mu_2 \neq 0$
    $\alpha = .05$
    C.R. $z < -z_{.025} = -1.960$
    $\quad\quad z > z_{.025} = 1.960$
    calculations:

    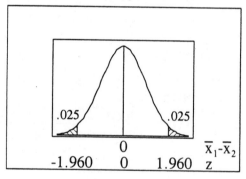

$$z_{\bar{x}_1 - \bar{x}_2} = (\bar{x}_1 - \bar{x}_2 - \mu_{\bar{x}_1 - \bar{x}_2})/\sigma_{\bar{x}_1 - \bar{x}_2}$$
$$= (1.00 - 0)/\sqrt{(1.00)^2/50 + (2.00)^2/100}$$
$$= 1.00/.2449 = 4.082$$

conclusion:
Reject H_0; there is sufficient evidence to reject the claim that $\mu_1 - \mu_2 = 0$ and to conclude that $\mu_1 - \mu_2 \neq 0$ (in fact, that $\mu_1 - \mu_2 > 0$).
P-value = $2 \cdot P(z > 4.08) = 2 \cdot (.5000 - .4999) = 2 \cdot (.0001) = .0002$

3. $(\bar{x}_1 - \bar{x}_2) \pm z_{.025}\sqrt{\sigma_1^2/n_1 + \sigma_2^2/n_2}$
    $1.00 \pm 1.960 \cdot \sqrt{(1.00)^2/50 + (2.00)^2/100}$
    $1.00 \pm .48$
    $.52 < \mu_1 - \mu_2 < 1.48$
    No; the confidence interval does not include 0. Since the confidence interval limits do not include 0, conclude that there is a real difference between the treatment and placebo groups.

5. a. original claim: $\mu_1-\mu_2 = 0$ [$n_1 > 30$ and $n_2 > 30$, use z (with s's for σ's)]
$\bar{x}_1-\bar{x}_2 = .81682 - .78479 = .03203$
H_o: $\mu_1-\mu_2 = 0$
H_1: $\mu_1-\mu_2 \neq 0$
$\alpha = .01$
C.R. $z < -z_{.005} = -2.575$
$\quad z > z_{.005} = 2.575$
calculations:

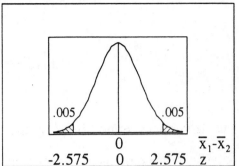

$z_{\bar{x}_1-\bar{x}_2} = (\bar{x}_1-\bar{x}_2 - \mu_{\bar{x}_1-\bar{x}_2})/\sigma_{\bar{x}_1-\bar{x}_2}$
$\quad = (.03203 - 0)/\sqrt{(.007507)^2/36 + (.004391)^2/36}$
$\quad = .03203/.0014495 = 22.098$

conclusion:
Reject H_o; there is sufficient evidence to reject the claim that $\mu_1-\mu_2 = 0$ and to conclude that $\mu_1-\mu_2 \neq 0$ (in fact, that $\mu_1-\mu_2 > 0$).
P-value = $2 \cdot P(z > 22.10) = 2 \cdot (.5000 - .4999) = 2 \cdot (.0001) = .0002$

b. $(\bar{x}_1-\bar{x}_2) \pm z_{.005}\sqrt{\sigma_1^2/n_1 + \sigma_2^2/n_2}$
$.03203 \pm 2.575 \cdot \sqrt{(.007507)^2/36 + (.004391)^2/36}$
$.03203 \pm .00373$
$.02830 < \mu_1-\mu_2 < .03576$

7. original claim: $\mu_1-\mu_2 > 0$ [$n_1 > 30$ and $n_2 > 30$, use z (with s's for σ's)]
$\bar{x}_1-\bar{x}_2 = 53.3 - 45.3 = 8.0$
H_o: $\mu_1-\mu_2 \leq 0$
H_1: $\mu_1-\mu_2 > 0$
$\alpha = .01$
C.R. $z > z_{.01} = 2.327$
calculations:

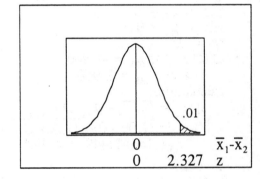

$z_{\bar{x}_1-\bar{x}_2} = (\bar{x}_1-\bar{x}_2 - \mu_{\bar{x}_1-\bar{x}_2})/\sigma_{\bar{x}_1-\bar{x}_2}$
$\quad = (8.0 - 0)/\sqrt{(11.6)^2/40 + (13.2)^2/40}$
$\quad = 8.0/2.778 = 2.879$

conclusion:
Reject H_o; there is sufficient evidence to conclude that $\mu_1-\mu_2 > 0$.
P-value = $P(z > 2.88) = .5000 - .4980 = .0020$

9. Let the younger men be group 1.
a. original claim: $\mu_1-\mu_2 > 0$ [$n_1 > 30$ and $n_2 > 30$, use z (with s's for σ's)]
$\bar{x}_1-\bar{x}_2 = 176 - 164 = 12$
H_o: $\mu_1-\mu_2 \leq 0$
H_1: $\mu_1-\mu_2 > 0$
$\alpha = .01$
C.R. $z > z_{.01} = 2.327$
calculations:

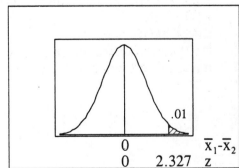

$z_{\bar{x}_1-\bar{x}_2} = (\bar{x}_1-\bar{x}_2 - \mu_{\bar{x}_1-\bar{x}_2})/\sigma_{\bar{x}_1-\bar{x}_2}$
$\quad = (12 - 0)/\sqrt{(35.0)^2/804 + (27.0)^2/1657}$
$\quad = 12/1.401 = 8.563$

conclusion:
Reject H_o; there is sufficient evidence to conclude that $\mu_1-\mu_2 > 0$.
P-value = $P(z > 8.56) = .5000 - .4999 = .0001$

b. $(\bar{x}_1 - \bar{x}_2) \pm z_{.005}\sqrt{\sigma_1^2/n_1 + \sigma_2^2/n_2}$

$12 \pm 2.575 \cdot \sqrt{(35.0)^2/804 + (27.0)^2/1657}$

12 ± 3.6

$8 < \mu_1 - \mu_2 < 16$

The confidence interval does not include the value 0. This indicates that there is a significant difference between the two means and agrees with the conclusion in part (a).

11. original claim: $\mu_1 - \mu_2 > 0$ [$n_1 > 30$ and $n_2 > 30$, use z (with s's for σ's)]

$\bar{x}_1 - \bar{x}_2 = 3214 - 3088 = 126$

$H_o: \mu_1 - \mu_2 \le 0$

$H_1: \mu_1 - \mu_2 > 0$

$\alpha = .05$

C.R. $z > z_{.05} = 1.645$

calculations:

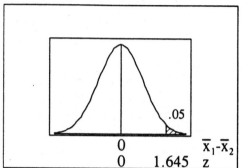

$z_{\bar{x}_1 - \bar{x}_2} = (\bar{x}_1 - \bar{x}_2 - \mu_{\bar{x}_1 - \bar{x}_2})/\sigma_{\bar{x}_1 - \bar{x}_2}$

$\quad = (126 - 0)/\sqrt{(669)^2/294 + (728)^2/286}$

$\quad = 126/58.1 = 2.169$

conclusion:

Reject H_o; there is sufficient evidence to conclude that $\mu_1 - \mu_2 > 0$.

P-value $= P(z > 2.17) = .5000 - .4850 = .0150$

13. let the University of Massachusetts new texts be group 1

group 1: UMASS (n = 40)   group 2: DCC (n = 35)

$\Sigma x = 2604.70$ $\Sigma x = 2016.85$

$\Sigma x^2 = 190378.5648$ $\Sigma x^2 = 123549.5728$

$\bar{x} = 65.12$ $\bar{x} = 57.62$

$s^2 = 532.487$ $s^2 = 215.589$

original claim: $\mu_1 - \mu_2 = 0$ [$n_1 > 30$ and $n_2 > 30$, use z (with s's for σ's)]

$\bar{x}_1 - \bar{x}_2 = 65.12 - 57.62 = 7.493$

$H_o: \mu_1 - \mu_2 = 0$

$H_1: \mu_1 - \mu_2 \ne 0$

$\alpha = .05$ [assumed]

C.R. $z < -z_{.025} = -1.960$

$\quad z > z_{.025} = 1.960$

calculations:

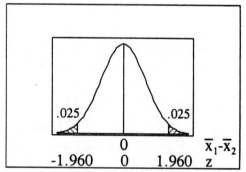

$z_{\bar{x}_1 - \bar{x}_2} = (\bar{x}_1 - \bar{x}_2 - \mu_{\bar{x}_1 - \bar{x}_2})/\sigma_{\bar{x}_1 - \bar{x}_2}$

$\quad = (7.493 - 0)/\sqrt{532.487/40 + 215.589/35}$

$\quad = 7.493/4.413 = 1.698$

conclusion:

Do not reject H_o; there is not sufficient evidence to reject the claim that $\mu_1 - \mu_2 = 0$.

P-value $= 2 \cdot P(z > 1.70) = 2 \cdot (.5000 - .4554) = 2 \cdot (.0446) = .0892$

15. let the non-smokers not exposed to second-hand smoke (NOETS) be group 1

group 1: NOETS (n = 50)   group 2: ETS (n = 50)

$\Sigma x = 20.25$ $\Sigma x = 204.88$

$\Sigma x^2 = 80.2869$ $\Sigma x^2 = 5949.2360$

$\bar{x} = .4050$ $\bar{x} = 4.0976$

$s^2 = 1.471$ $s^2 = 104.280$

original claim: $\mu_1 - \mu_2 = 0$ [$n_1 > 30$ and $n_2 > 30$, use z (with s's for σ's)]
$\bar{x}_1 - \bar{x}_2 = .4050 - 4.0976 = -3.6926$
H_0: $\mu_1 - \mu_2 = 0$
H_1: $\mu_1 - \mu_2 \neq 0$
$\alpha = .05$ [assumed]
C.R. $z < -z_{.025} = -1.960$
$z > z_{.025} = 1.960$
calculations:
$z_{\bar{x}_1 - \bar{x}_2} = (\bar{x}_1 - \bar{x}_2 - \mu_{\bar{x}_1 - \bar{x}_2})/\sigma_{\bar{x}_1 - \bar{x}_2}$
$= (-3.6926 - 0)/\sqrt{1.471/50 + 104.280/50}$
$= -3.6926/1.454 = -2.539$

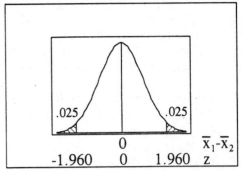

conclusion:
   Reject H_0; there is sufficient evidence to reject the claim that $\mu_1 - \mu_2 = 0$ and to conclude that $\mu_1 - \mu_2 \neq 0$ (in fact, that $\mu_1 - \mu_2 < 0$ -- i.e., that non-smokers not exposed to second-hand smoke have significantly less cotinine in their system than non-smokers so exposed).
P-value $= 2 \cdot P(z < -2.54) = 2 \cdot (.5000 - .4945) = 2 \cdot (.0055) = .0110$

17. original claim: $\mu_1 - \mu_2 < 0$ [$n_1 > 30$ and $n_2 > 30$, use z (with s's for σ's)]
$\bar{x}_1 - \bar{x}_2 = 152.0739 - 154.9669 = -2.893$
H_0: $\mu_1 - \mu_2 \geq 0$
H_1: $\mu_1 - \mu_2 < 0$
$\alpha = .05$
C.R. $z < -z_{.05} = -1.645$
calculations:
$z_{\bar{x}_1 - \bar{x}_2} = (\bar{x}_1 - \bar{x}_2 - \mu_{\bar{x}_1 - \bar{x}_2})/\sigma_{\bar{x}_1 - \bar{x}_2}$
$= (-2.893 - 0)/\sqrt{438.5388/50 + 239.1461/50}$
$= -.7858$ [from Excel]

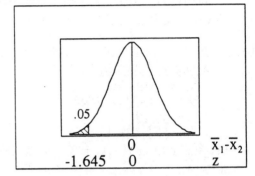

conclusion:
   Do not reject H_0; there is no sufficient evidence to conclude that $\mu_1 - \mu_2 < 0$. While the sample mean was less for treatment group, it is not unlike that this could have occurred by chance and there was not enough evidence to conclude that this pattern applies in the populations.
P-value $= P(z < -.7858) = .215991$ [from Excel]

19. original claim: $\mu_1 - \mu_2 > 0$ [$n_1 > 30$ and $n_2 > 30$, use z (with s's for σ's)]
$\bar{x}_1 - \bar{x}_2 = 68.898 - 63.862 = 5.036$
H_0: $\mu_1 - \mu_2 \leq 0$
H_1: $\mu_1 - \mu_2 > 0$
$\alpha = .05$ [assumed]
C.R. $z > z_{.05} = 1.645$
calculations:
$z_{\bar{x}_1 - \bar{x}_2} = (\bar{x}_1 - \bar{x}_2 - \mu_{\bar{x}_1 - \bar{x}_2})/\sigma_{\bar{x}_1 - \bar{x}_2}$
$= 9.964$ [from TI-83 Plus]

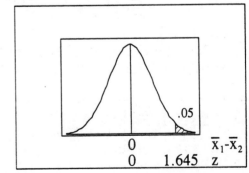

conclusion:
   Reject H_0; there is sufficient evidence to conclude that $\mu_1 - \mu_2 > 0$. The probability of obtaining our sample result if the mean height for men was not greater than the mean height for women is so small, that we conclude as indicated.
P-value $= p(z > 9.964) = 1.11E-23 = .00000000000000000000000111$ [from TI-83 Plus]

21. FIRST: use the data sets as given
let the .0111 in. cans be group 1

group 1: .0111 in. (n = 175)
$\sum x = 49316$
$\sum x^2 = 14031760$
$\bar{x} = 281.8$
$s^2 = 771.43$

group 2: .0109 in (n = 175)
$\sum x = 46745$
$\sum x^2 = 12571335$
$\bar{x} = 267.1$
$s^2 = 488.95$

original claim: $\mu_1-\mu_2 > 0$ [$n_1 > 30$ and $n_2 > 30$, use z (with s's for σ's)]
$\bar{x}_1-\bar{x}_2 = 281.8 - 267.1 = 14.7$
H_o: $\mu_1-\mu_2 \le 0$
H_1: $\mu_1-\mu_2 > 0$
$\alpha = .05$ [assumed]
C.R. $z > z_{.05} = 1.645$
calculations:

$$z_{\bar{x}_1-\bar{x}_2} = (\bar{x}_1-\bar{x}_2 - \mu_{\bar{x}_1-\bar{x}_2})/\sigma_{\bar{x}_1-\bar{x}_2}$$
$$= (14.7 - 0)/\sqrt{771.43/175 + 488.95/175}$$
$$= 14.7/2.684 = 5.474$$

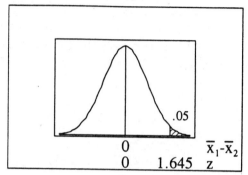

conclusion:
Reject H_o; there is sufficient evidence to conclude that $\mu_1-\mu_2 > 0$.
P-value = $P(z > 5.47) = .5000 - .4999 = .0001$

NEXT: consider the 504 to be an outlier and remove it from group 1

group 1: .0111 in. (n = 174)
$\sum x = 48812$
$\sum x^2 = 13777744$
$\bar{x} = 280.5$
$s^2 = 488.87$

group 2: .0109 in (n = 175)
$\sum x = 46745$
$\sum x^2 = 12571335$
$\bar{x} = 267.1$
$s^2 = 488.95$

NOTE: The new mean for group 1 is slightly smaller than the old one, and the new standard deviation is noticeably smaller -- just the types of changes one should expect when removing a large outlier.

original claim: $\mu_1-\mu_2 > 0$ [$n_1 > 30$ and $n_2 > 30$, use z (with s's for σ's)]
$\bar{x}_1-\bar{x}_2 = 280.5 - 267.1 = 13.4$
H_o: $\mu_1-\mu_2 \le 0$
H_1: $\mu_1-\mu_2 > 0$
$\alpha = .05$ [assumed]
C.R. $z > z_{.05} = 1.645$
calculations:

$$z_{\bar{x}_1-\bar{x}_2} = (\bar{x}_1-\bar{x}_2 - \mu_{\bar{x}_1-\bar{x}_2})/\sigma_{\bar{x}_1-\bar{x}_2}$$
$$= (13.4 - 0)/\sqrt{488.87/174 + 488.95/175}$$
$$= 13.4/2.367 = 5.661$$

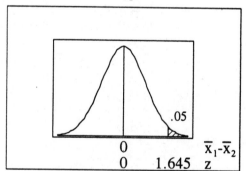

conclusion:
Reject H_o; there is sufficient evidence to conclude that $\mu_1-\mu_2 > 0$.
P-value = $P(z > 5.66) = .5000 - .4999 = .0001$
NOTE: Even though the difference between the means was smaller, the decreased variability actually resulted in a slightly higher calculated z and a more significant difference.

ALSO: answer the questions about boxplots and confidence intervals

boxplots:

old 5-number boxplot summary (group 1)

$L = x_1 = 205$
$.25(175) = 43.74; P_{25} = x_{44} = 275$
$.50(175) = 87.50; P_{50} = x_{88} = 285$

$.75(175) = 131.25; P_{75} = x_{132} = 295$
$H = x_{175} = 504$

new 5-number boxplot summary (group 1)

$L = x_1 = 205$
$.25(174) = 43.5; P_{25} = x_{44} = 275$
$.50(174) = 87; P_{50} = (x_{87}+x_{88})/2$
$\qquad\qquad = (285+285)/2 = 285$
$.75(174) = 130.5; P_{75} = x_{131} = 294$
$H = x_{174} = 317$

The new boxplot is given below. The only significant difference is the smaller length of the extension from the third quartile to the maximum value, which would extend all the way to 504 for the original data.

205   275 285 294   317

confidence intervals:

original 95% confidence interval ($\mu_1-\mu_2$)

$(\bar{x}_1-\bar{x}_2) \pm z_{.025}\sqrt{\sigma_1^2/n_1 + \sigma_2^2/n_2}$
$(281.8 - 267.1) \pm 1.960 \cdot 2.684$

14.7 ± 5.3
$9.4 < \mu_1-\mu_2 < 20.0$

new 95% confidence interval ($\mu_1-\mu_2$)

$(\bar{x}_1-\bar{x}_2) \pm z_{.025}\sqrt{\sigma_1^2/n_1 + \sigma_2^2/n_2}$
$(280.5 - 267.1) \pm 1.960 \cdot 2.367$

13.4 ± 4.6
$8.8 < \mu_1-\mu_2 < 18.0$

The decrease in variability realized by dropping the outlier created a slightly narrower confidence interval.

CONCLUSION: In general, this sample is so large that the single outlier does not have a major effect on the test of hypothesis or the confidence interval for the difference between the means. The only major effect the outlier has is on the spread (as seen in the standard deviation and boxplot) of group 1.

23. a. $x = 5,10,15$
$\quad \mu = \Sigma x/n = 30/3 = 10$
$\quad \sigma^2 = \Sigma(x-\mu)^2/n$
$\qquad = [(-5)^2 + (0)^2 + (5)^2]/3$
$\qquad = 50/3$

b. $y = 1,2,3$
$\quad \mu = \Sigma y/n = 6/3 = 2$
$\quad \sigma^2 = \Sigma(y-\mu)^2/n$
$\qquad = [(-1)^2 + (0)^2 + (1)^2]/3$
$\qquad = 2/3$

c. $z = x-y = 4,3,2,9,8,7,14,13,12$
$\quad \mu = \Sigma z/n = 72/9 = 8$
$\quad \sigma^2 = \Sigma(z-\mu)^2/n$
$\qquad = [(-4)^2 + (-5)^2 + (-6)^2 + (1)^2 + (0)^2 + (-1)^2 + (6)^2 + (5)^2 + (4)^2]/9$
$\qquad = 156/9$
$\qquad = 52/3$

d. $\sigma_{x-y}^2 = \sigma_x^2 + \sigma_y^2$
$\quad 52/3 = 50/3 + 2/3$
$\quad 52/3 = 52/3$

e. Let R stand for range.
$\quad R_{x-y} = \text{highest}_{x-y} - \text{lowest}_{x-y}$
$\qquad = (\text{highest } x - \text{lowest } y) - (\text{lowest } x - \text{highest } y)$
$\qquad = \text{highest } x - \text{lowest } y - \text{lowest } x + \text{highest } y$
$\qquad = (\text{highest } x - \text{lowest } x) + (\text{highest } y - \text{lowest } y)$
$\qquad = R_x + R_y$
The range of all possible x-y values is the sum of the individual ranges of x and y.
NOTE: The problem refers to all possible x-y differences (where n_x and n_y might even be different) and not to x-y differences for paired data.

8-3 Inferences about Two Means: Matched Pairs

NOTE: To be consistent with the notation of the previous sections, and thereby reinforcing the patterns and concepts presented in those sections, the manual uses the "usual" t formula written to apply to d's

$$t_{\bar{d}} = (\bar{d} - \mu_{\bar{d}})/s_{\bar{d}}$$

with $\mu_{\bar{d}} = \mu_d$

and $s_{\bar{d}} = s_d/\sqrt{n}$

And so the formula for the t statistic may also be written as $t_{\bar{d}} = (\bar{d} - \mu_d)/(s_d/\sqrt{n})$

1. d = x - y: 5 6 0 5 -2
   summary statistics: n = 5, $\Sigma d = 14$, $\Sigma d^2 = 90$
   a. $\bar{d} = (\Sigma d)/n = 14/5 = 2.8$
   b. $s_d^2 = [n \cdot \Sigma d^2 - (\Sigma d)^2]/[n(n-1)]$
      $= [5 \cdot 90 - (14)^2]/[5(4)]$
      $= 254/20 = 12.7$
      $s_d = 3.6$
   c. $t_{\bar{d}} = (\bar{d} - \mu_{\bar{d}})/s_{\bar{d}}$
      $= (2.8 - 0)/(3.564/\sqrt{5})$
      $= 2.8/1.594 = 1.757$
   d. $\pm t_{4,.025} = \pm 2.776$

3. $\bar{d} \pm t_{4,.025} \cdot s_d/\sqrt{n}$
   $2.8 \pm 2.776 \cdot 3.564/\sqrt{5}$
   2.8 ± 4.4
   $-1.6 < \mu_d < 7.2$

5. d = x - y: 0.5 -0.1 0.2 1.6 1.9 -0.4 0.3 -0.3 0.6 1.8
   n = 10
   $\Sigma d = 6.1$ $\bar{d} = .61$
   $\Sigma d^2 = 10.41$ $s_d = .862$
   a. original claim: $\mu_d > 0$ [n ≤ 30 and σ_d unknown, use t]
      $H_0: \mu_d \leq 0$
      $H_1: \mu_d > 0$
      $\alpha = .05$
      C.R. $t > t_{9,.05} = 1.833$
      calculations:
         $t_{\bar{d}} = (\bar{d} - \mu_{\bar{d}})/s_{\bar{d}}$
         $= (.61 - 0)/(.862/\sqrt{10})$
         $= .61/.273 = 2.238$

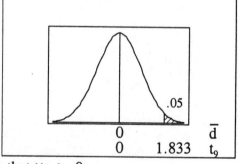

   conclusion:
      Reject H_0; there is sufficient evidence to conclude that $\mu_d > 0$.
   b. $\bar{d} \pm t_{9,.025} \cdot s_d/\sqrt{n}$
      $.61 \pm 2.262 \cdot .862/\sqrt{10}$
      $.61 \pm .62$
      $-.01 < \mu_d < 1.23$
   Since the confidence interval contains 0, there is no significant difference between the reported and measured heights. This does not contradict the result in part (a). Since the test of hypothesis was one-tailed test and the confidence interval is two-tailed, they are not equivalent. Together they show the value of one-tailed tests: a one-tailed test (remember: all claims must be made before seeing the data) can result in the rejection of the null hypothesis when a two-tailed test at the same level would not.

7. d = x - y: -20 0 10 -40 -30 -10 30 -20 -20 -10
    n = 10
    Σd = -110 $\bar{d}$ = -11.0
    Σd^2 = 4900 s_d = 20.248
    a. original claim: $\mu_d < 0$ [n $\leq$ 30 and σ_d unknown, use t]
        H_o: $\mu_d \geq 0$
        H_1: $\mu_d < 0$
        α = .05
        C.R. t < $-t_{9,.05}$ = -1.833
        calculations:
            $t_{\bar{d}}$ = $(\bar{d} - \mu_{\bar{d}})/s_{\bar{d}}$
                = (-11.0 - 0)/(20.248/$\sqrt{10}$)
                = -11.0/6.40 = -1.718

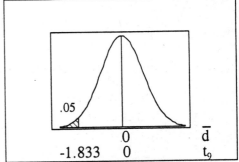

    conclusion:
        Do not reject H_o; there is not sufficient evidence to conclude that $\mu_d < 0$.
    b. $\bar{d} \pm t_{9,.025} \cdot s_d/\sqrt{n}$
        -11.0 $\pm$ 2.262·20.248/$\sqrt{10}$
        -11.0 $\pm$ 14.5
        -25.5 < μ_d < 3.5
    We have 95% confidence that the interval from -25.5 to 3.5 contains the true mean
    population difference. Since this interval includes 0, the mean before and after scores
    are not significantly different, and there is not enough evidence to say that the course
    has any effect.

9. d = $x_B - x_A$: -.2 4.1 1.6 1.8 3.2 2.0 2.9 9.6
    n = 8
    Σd = 25.0 $\bar{d}$ = 3.125
    Σd^2 = 137.46 s_d = 2.9114
    a. asks for confidence interval for μ_d [n $\leq$ 30 and σ_d unknown, use t]
        $\bar{d} \pm t_{7,.025} \cdot s_d/\sqrt{n}$
        3.125 $\pm$ 2.365·2.9114/$\sqrt{8}$
        3.125 $\pm$ 2.434
        .69 < μ_d < 5.56
    b. original claim: $\mu_d > 0$
        H_o: $\mu_d \leq 0$
        H_1: $\mu_d > 0$
        α = .05
        C.R. t > $t_{7,.05}$ = 1.895
        calculations:
            $t_{\bar{d}}$ = $(\bar{d} - \mu_{\bar{d}})/s_{\bar{d}}$
                = (3.125 - 0)/(2.9114/$\sqrt{8}$)
                = 3.125/1.029 = 3.036

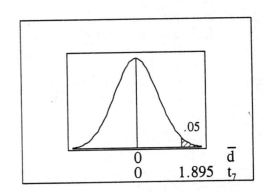

    conclusion:
        Reject H_o; there is sufficient evidence to conclude that $\mu_d > 0$.
    c. Yes; hypnotism appears to be effective in reducing pain.

11. $d = x_8 - x_{12}$: given at the right; asks about μ_d [$n \leq 30$ and σ_d unknown, use t]

sbj#	8am	12am	diff
19	97.0	97.7	-0.7
20	98.0	98.8	-0.8
22	96.4	98.0	-1.6
26	98.2	98.7	-0.5
71	98.8	98.0	0.8
78	98.6	98.5	0.1
80	97.8	98.3	-0.5
81	98.7	98.7	0.0
83	97.8	99.1	-1.3
98	96.4	98.2	-1.8
99	96.9	99.2	-2.3

$n = 11$
$\Sigma d = -8.6$ $\overline{d} = -.782$
$\Sigma d^2 = 15.06$ $s_d = .913$

a. $\overline{d} \pm t_{10,.025} \cdot s_d/\sqrt{n}$
   $-.782 \pm 2.228 \cdot .913/\sqrt{11}$
   $-.782 \pm .613$
   $-1.40 < \mu_d < -.17$

b. original claim: $\mu_d = 0$
   $H_o: \mu_d = 0$
   $H_1: \mu_d \neq 0$
   $\alpha = .05$
   C.R. $t < -t_{10,.025} = -2.228$
   $\quad t > t_{10,.025} = 2.228$
   calculations:
   $t_{\overline{d}} = (\overline{d} - \mu_{\overline{d}})/s_{\overline{d}}$
   $\quad = (-.782 - 0)/(.913/\sqrt{11})$
   $\quad = -.782/.275 = -2.840$

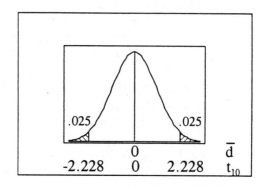

conclusion:
   Reject H_o; there is sufficient evidence to reject the claim that $\mu_d = 0$ and conclude
   that $\mu_d \neq 0$ (in fact, that $\mu_d < 0$.
No; based on this result. morning and night body temperatures do not appear to be
about the same. The 8 am (morning) temperatures are significantly lower.

13. $d = x_B - x_A$; asks about μ_d [$n \leq 30$ and σ_d unknown, use t]

a. original claim: $\mu_d \neq 0$
   $H_o: \mu_d = 0$
   $H_1: \mu_d \neq 0$
   $\alpha = .05$
   C.R. $t < -t_{9,.025} = -2.262$
   $\quad t > t_{9,.025} = 2.262$
   calculations:
   $t_{\overline{d}} = (\overline{d} - \mu_{\overline{d}})/s_{\overline{d}}$
   $\quad = (-7.5 - 0)/(57.7/\sqrt{10})$
   $\quad = -7.5/18.2 = -.41$ [from Minitab]

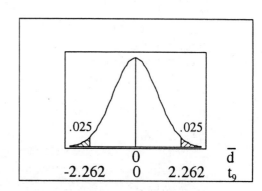

conclusion:
   Do not reject H_o; there is not sufficient evidence to conclude that $\mu_d \neq 0$.
No; based on this result, do not spend the money for the drug. NOTE: The conclusion
could also have been reached by the P-value method. Since the Minitab's P-value =
$2 \cdot P(t_9 < -.41) = .69$ is greater than $\alpha = .05$, there is not sufficient evidence to reject
H_o.

b. original claim: $\mu_d < 0$

$\quad$ H_o: $\mu_d \geq 0$

$\quad$ H_1: $\mu_d < 0$

$\quad$ $\alpha = .05$

$\quad$ C.R. $t < -t_{9,.05} = -1.833$

$\quad$ calculations:

$\qquad$ $t_{\bar{d}} = (\bar{d} - \mu_{\bar{d}})/s_{\bar{d}}$

$\qquad\quad$ $= (-7.5 - 0)/(57.7/\sqrt{10})$

$\qquad\quad$ $= -7.5/18.2 = -.41$ [from Minitab]

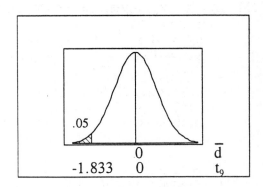

conclusion:

$\quad$ Do not reject H_o; there is not sufficient evidence to conclude that $\mu_d < 0$.

The calculations and the conclusion are the same as in part (a), but now the P-value = $P(t_9 < -.41) = \frac{1}{2}(.69) = .345$, which is still greater than $\alpha = .05$.

c. $\bar{d} \pm t_{9,.025} \cdot s_d/\sqrt{n}$

$\quad$ $-7.5 \pm 2.262 \cdot 57.7/\sqrt{10}$

$\quad$ -7.5 ± 41.3

$\quad$ $-48.8 < \mu_d < 33.8$

Yes; the confidence interval includes the value 0, indicating that the drug does not have a significant effect on vulnerability to motion sickness as measured.

15. summary statistics:

$\quad$ $n = 12$

$\quad$ $\Sigma d = 15.35$ $\qquad\qquad$ $\bar{d} = 1.279$

$\quad$ $\Sigma d^2 = 75.0025$ $\qquad\quad$ $s_d = 2.244$

hypothesis test for the original claim: $\mu_d > 0$ [$n \leq 30$ and σ_d unknown, use t]

$\quad$ H_o: $\mu_d \leq 0$

$\quad$ H_1: $\mu_d > 0$

$\quad$ $\alpha = .05$

$\quad$ C.R. $t > t_{11,.05} = 1.796$

$\quad$ calculations:

$\qquad$ $t_{\bar{d}} = (\bar{d} - \mu_{\bar{d}})/s_{\bar{d}}$

$\qquad\quad$ $= (1.279 - 0)/(2.244/\sqrt{12})$

$\qquad\quad$ $= 1.279/.6476 = 1.975$

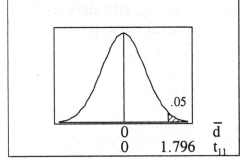

conclusion:

$\quad$ Reject H_o; there is sufficient evidence to conclude that $\mu_d > 0$.

While the overall conclusion is the same, including the outlier changed the test considerably. The point estimate of μ_d almost doubled from .673 to 1.279, and the point estimate of σ_d almost tripled from .826 to 2.244. The increase in the estimate of the difference μ_d (which increases the significance of the sample result as measured by the calculated t) was offset by the increase in the estimate of the variability σ_d (which decreases the significance of the sample result as measured by the calculated t) and so the test statistic decreased from 2.701 to 1.975.

confidence interval:

$\quad$ $\bar{d} \pm t_{11,.025} \cdot s_d/\sqrt{n}$

$\quad$ $1.279 \pm 2.201 \cdot 2.244/\sqrt{12}$

$\quad$ 1.279 ± 1.425

$\quad$ $-.146 < \mu_d < 2.705$

Even though the point estimate without the outlier was closer to 0, the confidence interval did not include 0 and suggested that there was a significant positive bias in the reported heights. Even though the point estimate including the outlier was farther from 0, the increased variability made the interval wide enough to include the value 0.

Yes; in general, an outlier can have a dramatic effect on the hypothesis test and the confidence interval.

NOTE: In order to obtain the best analysis possible and to avoid future outliers, most practicing statisticians will try to determine why a particular outlier may have occurred. In this instance it is plausible that the person was reporting 6'2.25" (which would be 74.25, very close to the actual value of 74.3) but either (a) he mistakenly thought 6' translated into 80 inches and wrote 82.25" or (b) instead of recording the value in inches as directed, he wrote 6'2.25" and his handwritten "6'" was mistaken for "8" (a careful check of the original data often clarifies supposed outliers).

17. Since the 95% confidence interval $\overline{d} \pm t_{df,.025} \cdot s_d/\sqrt{n}$ is $0 < \mu_d < 1.2$, we know the following:

$$\overline{d} - t_{df,.025} \cdot s_d/\sqrt{n} = 0$$
$$\overline{d} = t_{df,.025} \cdot s_d/\sqrt{n}$$
$$\overline{d} - 0 = t_{df,.025} \cdot s_d/\sqrt{n}$$
$$(\overline{d} - 0)/(\cdot s_d/\sqrt{n}) = t_{df,.025}$$

Testing the claim $\mu_d > 0$ at the α level produces the following:

$H_o: \mu_d \leq 0$
$H_1: \mu_d > 0$
$\alpha = ?$
C.R. $t > t_{df,\alpha}$
calculations:

$$t_{\overline{d}} = (\overline{d} - \mu_{\overline{d}})/s_{\overline{d}}$$
$$= (\overline{d} - 0)/(\cdot s_d/\sqrt{n})$$
$$= t_{df,.025} \text{ [from the confidence interval above]}$$

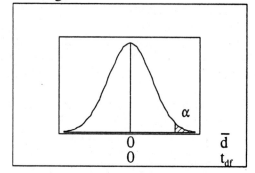

conclusion:
Reject H_o if the critical t is smaller than the calculated t -- i.e., whenever
$$t_{df,\alpha} < t_{df,.025}.$$

Since $t_{df,\alpha} < t_{df,.025}$ occurs when $\alpha > .025$, the smallest possible value of α leading to the stipulated rejection of H_o is $\alpha = .025$.

8-4 Inferences about Two Proportions

NOTE: To be consistent with the notation of the previous sections, thereby reinforcing the patterns and concepts presented there, the manual uses the "usual" z formula written to apply to $\hat{p}_1 - \hat{p}_2$'s

$$z_{\hat{p}_1 - \hat{p}_2} = (\hat{p}_1 - \hat{p}_2 - \mu_{\hat{p}_1 - \hat{p}_2})/\sigma_{\hat{p}_1 - \hat{p}_2}$$

with $\mu_{\hat{p}_1 - \hat{p}_2} = p_1 - p_2$
and $\sigma_{\hat{p}_1 - \hat{p}_2} = \sqrt{\overline{pq}/n_1 + \overline{pq}/n_2}$ [when H_o includes $p_1 = p_2$]
where $\overline{p} = (x_1 + x_2)/(n_1 + n_2)$
And so the formula for the z statistic may also be written as

$$z_{\hat{p}_1 - \hat{p}_2} = ((\hat{p}_1 - \hat{p}_2) - (p_1 - p_2))/\sqrt{\overline{pq}/n_1 + \overline{pq}/n_2}$$

1. $x = \hat{p} \cdot n = .0425 \cdot 2750 = 116.875$, rounded to 117

3. $x = \hat{p} \cdot n = .289 \cdot 294 = 84.966$, rounded to 85

5. $\hat{p}_1 = x_1/n_1 = 10/20 = .500$
$\hat{p}_2 = x_2/n_2 = 15/25 = .600$
$\hat{p}_1 - \hat{p}_2 = .500 - .600 = -.100$
   a. $\bar{p} = (x_1 + x_2)/(n_1 + n_2)$
   $= (10 + 15)/(20 + 25)$
   $= 25/45 = .556$
   b. $z_{\hat{p}_1-\hat{p}_2} = (\hat{p}_1-\hat{p}_2 - \mu_{\hat{p}_1-\hat{p}_2})/\sigma_{\hat{p}_1-\hat{p}_2}$
   $= (-.100 - 0)/\sqrt{(.556)(.444)/20 + (.556)(.444)/25}$
   $= -.100/.1491$
   $= -.671$
   c. $\pm z_{.025} = \pm 1.960$
   d. P-value $= 2 \cdot P(z < -.67) = 2 \cdot (.5000 - .2486) = 2 \cdot (.2514) = .5028$

NOTE: Since $\bar{p}$ is the weighted average of $\hat{p}_1$ and $\hat{p}_2$, it must always fall between those two values. If it does not, then an error has been made that must be corrected before proceeding. Calculation of $\sigma_{\hat{p}_1-\hat{p}_2} = \sqrt{\bar{p}\bar{q}/n_1 + \bar{p}\bar{q}/n_2}$ can be accomplished with no round-off loss on most calculators by calculating $\bar{p}$ and proceeding as follows: STORE 1-RECALL = * RECALL = STORE RECALL $\div$ n_1 + RECALL $\div$ n_2 = $\sqrt{}$. The quantity $\sigma_{\hat{p}_1-\hat{p}_2}$ may then be STORED for future use. Each calculator is different -- learn how your calculator works, and do the homework on the same calculator you will use for the exam. If you have any questions about performing/storing calculations on your calculator, check with your instructor or class assistant.

7. Let those who received the vaccine be group 1.
   original claim: $p_1 - p_2 < 0$
   $\hat{p}_1 = x_1/n_1 = 14/1070 = .013$
   $\hat{p}_2 = x_2/n_2 = 95/532 = .179$
   $\hat{p}_1 - \hat{p}_2 = .013 - .179 = -.165$
   $\bar{p} = (x_1 + x_2)/(n_1 + n_2)$
   $= (14 + 95)/(1070 + 532)$
   $= 109/1602 = .068$
   H_o: $p_1 - p_2 \geq 0$
   H_1: $p_1 - p_2 < 0$
   $\alpha = .05$ [assumed]
   C.R. $z < -z_{.05} = -1.645$

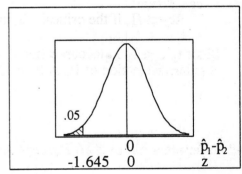

   calculations:
   $z_{\hat{p}_1-\hat{p}_2} = (\hat{p}_1-\hat{p}_2 - \mu_{\hat{p}_1-\hat{p}_2})/\sigma_{\hat{p}_1-\hat{p}_2}$
   $= (-.165 - 0)/\sqrt{(.068)(.932)/1070 + (.068)(.932)/532}$
   $= -.165/.0134 = -12.388$
   conclusion:
   Reject H_o; there is sufficient evidence to conclude that $p_1 - p_2 < 0$.
   P-value $= P(z < -12.39) = .5000 - .4999 = .0001$

9. Let those not wearing seat belts be group 1.
   original claim: $p_1-p_2 > 0$
   $\hat{p}_1 = x_1/n_1 = 50/290 = .1724$
   $\hat{p}_2 = x_2/n_2 = 16/123 = .1301$
   $\hat{p}_1-\hat{p}_2 = .1724 - .1301 = .0423$
   $\bar{p} = (x_1 + x_2)/(n_1 + n_2)$
   $\quad = (50 + 16)/(290 + 123)$
   $\quad = 66/413 = .160$
   H_o: $p_1-p_2 \leq 0$
   H_1: $p_1-p_2 > 0$
   $\alpha = .05$
   C.R. $z > z_{.05} = 1.645$

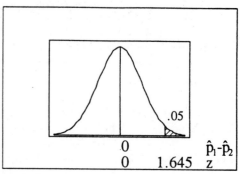

   calculations:
   $z_{\hat{p}1-\hat{p}2} = (\hat{p}_1-\hat{p}_2 - \mu_{\hat{p}1-\hat{p}2})/\sigma_{\hat{p}1-\hat{p}2}$
   $\quad = (.0423 - 0)/\sqrt{(.160)(.840)/290 + (.160)(.840)/123}$
   $\quad = .0423/.0394 = 1.074$
   conclusion:
       Do not reject H_o; there is not sufficient evidence to conclude that $p_1-p_2 > 0$.
   P-value = $P(z > 1.07) = .5000 - .3577 = .1423$
   Based on these results, no specific action should be taken.
   NOTE: This test involves only children who were hospitalized. It is probably true that a much greater percentage of children wearing seat belts avoided having to be hospitalized in the first place.

11. Let those receiving the Salk vaccine be group 1.
    original claim: $p_1-p_2 < 0$
    $\hat{p}_1 = x_1/n_1 = 33/200,000 = .000165$
    $\hat{p}_2 = x_2/n_2 = 115/200,000 = .000575$
    $\hat{p}_1-\hat{p}_2 = .000165 - .000575 = -.000410$
    $\bar{p} = (x_1 + x_2)/(n_1 + n_2)$
    $\quad = (33 + 115)/(200,000 + 200,000)$
    $\quad = 148/400,000 = .00037$
    H_o: $p_1-p_2 \geq 0$
    H_1: $p_1-p_2 < 0$
    $\alpha = .01$
    C.R. $z < -z_{.01} = -2.327$

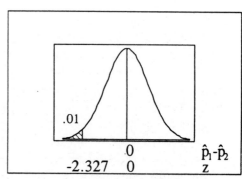

    calculations:
    $z_{\hat{p}1-\hat{p}2} = (\hat{p}_1-\hat{p}_2 - \mu_{\hat{p}1-\hat{p}2})/\sigma_{\hat{p}1-\hat{p}2}$
    $\quad = (-.000410 - 0)/\sqrt{(.00037)(.99963)/200,000 + (.00037)(.99963)/200,000}$
    $\quad = -.000410/.0000608 = -6.7416$ [from TI-83 Plus]
    conclusion:
        Reject H_o; there is sufficient evidence to conclude that $p_1-p_2 < 0$.
    P-value = $P(z < -6.74) = 7.88E-12 = .00000000000788$ [from TI-83 Plus]
    Yes; assuming the group assignments were made at random, the vaccine appears to be effective.
    NOTE: The values .00165, .00575 and .00037 appear above in the intermediate steps of the problem and in scientific notation on the TI-83 Plus. Even problems with calculator and/or software display given in the text will typically be worked out in some detail in this manual. For better understanding, match the numbers appearing in the displays with those appearing in the intermediate steps.

13. Let those younger than 21 be group 1.

$\hat{p}_1 = x_1/n_1 = .0425$

$\hat{p}_2 = x_2/n_2 = .0455$

$\hat{p}_1 - \hat{p}_2 = .0425 - .0455 = -.0029$

$(\hat{p}_1 - \hat{p}_2) \pm z_{.025}\sqrt{\hat{p}_1\hat{q}_1/n_1 + \hat{p}_2\hat{q}_2/n_2}$

$-.0029 \pm 1.960 \cdot \sqrt{(.0425)(.9575)/2750 + (.0455)(.9545)/2200}$

$-.0029 \pm .0115$

$-.0144 < p_1 - p_2 < .0086$

Yes; the interval contains zero, indicating no significant difference between the two rates of violent crimes.

15. Let those given the written survey be group 1.

$\hat{p}_1 = x_1/n_1 = 67/850 = .079 \ (x_1 = .079 \cdot 850)$

$\hat{p}_2 = x_2/n_2 = 105/850 = .124 \ (x_2 = .124 \cdot 850)$

$\hat{p}_1 - \hat{p}_2 = .079 - .124 = -.0447$

$\bar{p} = (x_1 + x_2)/(n_1 + n_2)$

$= (67 + 105)/(850 + 850)$

$= 172/1700 = .1012$

a. original claim: $p_1 - p_2 = 0$

$H_o: p_1 - p_2 = 0$

$H_1: p_1 - p_2 \neq 0$

$\alpha = .05$ [assumed]

C.R. $z < -z_{.025} = -1.960$

$\quad z > z_{.025} = 1.960$

calculations:

$z_{\hat{p}_1-\hat{p}_2} = (\hat{p}_1 - \hat{p}_2 - \mu_{\hat{p}_1-\hat{p}_2})/\sigma_{\hat{p}_1-\hat{p}_2}$

$= (-.0447 - 0)/\sqrt{(.1012)(.8988)/850 + (.1012)(.8988)/850}$

$= -.0447/.0146 = -3.056$

conclusion:

Reject H_o; there is sufficient evidence to reject the claim that $p_1 - p_2 = 0$ and to conclude

that $p_1 - p_2 \neq 0$ (in fact, that $p_1 - p_2 < 0$ -- i.e., that fewer students receiving the written test admit carrying a gun.)

P-value $= 2 \cdot P(z < -3.06) = 2 \cdot (.5000 - .4989) = 2 \cdot (.0011) = .0022$

Yes; based on this result the difference between 7.9% and 12.4% is significant.

b. $(\hat{p}_1 - \hat{p}_2) \pm z_{.005}\sqrt{\hat{p}_1\hat{q}_1/n_1 + \hat{p}_2\hat{q}_2/n_2}$

$-.0447 \pm 2.575 \cdot \sqrt{(.079)(.921)/850 + (.124)(.876)/850}$

$-.0447 \pm .0376$

$-.0823 < p_1 - p_2 < -.0071$

The interval does not contain zero, indicating a significant difference between the two response rates. We are 99% confident the interval from -8.3% to -0.7% contains the true difference between the population percentages.

17. Let the convicted arsonists be group 1.
    original claim: $p_1-p_2 > 0$
    $\hat{p}_1 = x_1/n_1 = 50/(50 + 43) = 50/93 = .538$
    $\hat{p}_2 = x_2/n_2 = 63/(63 + 144) = 63/207 = .304$
    $\hat{p}_1-\hat{p}_2 = .538 - .304 = .233$
    $\bar{p} = (x_1 + x_2)/(n_1 + n_2)$
    $\quad = (50 + 63)/(93 + 207)$
    $\quad = 113/300 = .377$
    H_o: $p_1-p_2 \le 0$
    H_1: $p_1-p_2 > 0$
    $\alpha = .01$
    C.R. $z > z_{.01} = 2.327$

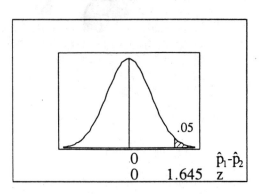

calculations:
$z_{\hat{p}1-\hat{p}2} = (\hat{p}_1-\hat{p}_2 - \mu_{\hat{p}1-\hat{p}2})/\sigma_{\hat{p}1-\hat{p}2}$
$\quad = (.233 - 0)/\sqrt{(.377)(.623)/93 + (.377)(.623)/207}$
$\quad = .233/.0605 = 3.857$
conclusion:
   Reject H_o; there is sufficient evidence to conclude that $p_1-p_2 > 0$.
P-value = $P(z > 3.86) = .5000 - .4999 = .0001$
Yes; it does seem reasonable that drinking might be related to the type of crime. The kinds of problems and personalities associated with drinking may well be more likely to be associated with some crimes more than others.

19. Let those receiving warmth be group 1.
    original claim: $p_1-p_2 < 0$
    $\hat{p}_1 = x_1/n_1 = 6/104 = .0577$
    $\hat{p}_2 = x_2/n_2 = 18/96 = .1875$
    $\hat{p}_1-\hat{p}_2 = .0577 - .1875 = -.1298$
    $\bar{p} = (x_1 + x_2)/(n_1 + n_2)$
    $\quad = (6 + 18)/(104 + 96)$
    $\quad = 24/200 = .12$
    H_o: $p_1-p_2 \ge 0$
    H_1: $p_1-p_2 < 0$
    $\alpha = .05$
    C.R. $z < -z_{.05} = -1.645$

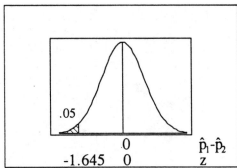

calculations:
$z_{\hat{p}1-\hat{p}2} = (\hat{p}_1-\hat{p}_2 - \mu_{\hat{p}1-\hat{p}2})/\sigma_{\hat{p}1-\hat{p}2}$
$\quad = (-.1298 - 0)/\sqrt{(.12)(.88)/104 + (.12)(.88)/96}$
$\quad = -.1298/.0460 = -2.822$
conclusion:
   Reject H_o; there is sufficient evidence to conclude that $p_1-p_2 < 0$.
P-value = $P(z < -2.82) = .5000 - .4976 = .0024$
Yes; if these results are verified, surgical patients should be routinely warmed.

21. Let those using Viagra be group 1.

original claim: $p_1-p_2 = .10$

$\hat{p}_1 = x_1/n_1 = .16$ [see note below]

$\hat{p}_2 = x_2/n_2 = .04$ [see note below]

$\hat{p}_1-\hat{p}_2 = .16 - .04 = .12$

H_o: $p_1-p_2 = .10$

H_1: $p_1-p_2 \neq .10$

$\alpha = .05$

C.R. $z < -z_{.025} = -1.960$

$z > z_{.025} = 1.960$

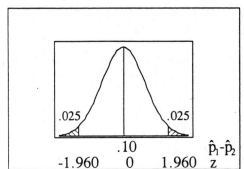

.025   .025

.10 $\hat{p}_1-\hat{p}_2$

-1.960   0   1.960   z

calculations:

$z_{\hat{p}1-\hat{p}2} = (\hat{p}_1-\hat{p}_2 - \mu_{\hat{p}1-\hat{p}2})/\sigma_{\hat{p}1-\hat{p}2}$

$= (.12 - .10)/\sqrt{(.16)(.84)/734 + (.04)(.96)/725}$

$= .02/.0154 = 1.302$

conclusion:

Do not reject H_o; there is not sufficient evidence to reject the claim that $p_1-p_2 = .10$.

P-value $= 2 \cdot P(z > 1.30) = 2 \cdot (.5000 - .4032) = 2 \cdot (.0968) = .1936$

NOTE: The values of x_1 and x_2 were not given. Since any x_1 between 114 and 121 inclusive and any x_2 between 26 and 32 inclusive round to the stated percents of 16% and 4%, no further accuracy is possible. The calculations proceed using only whole percents, and any attempt to identify more significant digits cannot be justified from the statement of the problem.

Review Exercises

1. Let those who took Viagra be group 1.

NOTE: Since there are a range of x_1 values and x_2 values that give $\hat{p}$'s which round to .07 and .02, and the original data is not given, the accuracy of this problem is limited. Use .07 and .02 when $\hat{p}$'s are called for. Use $x = \hat{p} \cdot n$ when x's are called for, even if that x is not a whole number. No other accuracy can justified from the given information.

a. original claim: $p_1-p_2 > 0$

$\hat{p}_1 = x_1/n_1 = 51.4/734 = .07$ [$x_1 = .07 \cdot 734 = 51.4$]

$\hat{p}_2 = x_2/n_2 = 14.5/725 = .02$ [$x_2 = .02 \cdot 725 = 14.5$]

$\hat{p}_1-\hat{p}_2 = .07 - .02 = .05$

$\bar{p} = (x_1 + x_2)/(n_1 + n_2)$

$= (51.4 + 14.5)/(734 + 725)$

$= 65.9/1459 = .04515$

H_o: $p_1-p_2 \leq 0$

H_1: $p_1-p_2 > 0$

$\alpha = .05$ [assumed]

C.R. $z > z_{.05} = 1.645$

.05

0   $\hat{p}_1-\hat{p}_2$

0   1.645   z

calculations:

$z_{\hat{p}1-\hat{p}2} = (\hat{p}_1-\hat{p}_2 - \mu_{\hat{p}1-\hat{p}2})/\sigma_{\hat{p}1-\hat{p}2}$

$= (.05 - 0)/\sqrt{(.04515)(.95485)/734 + (.04515)(.95485)/725}$

$= .05/.01087 = 4.599$

conclusion:

Reject H_o; there is sufficient evidence to conclude that $p_1-p_2 > 0$.

b. $(\hat{p}_1-\hat{p}_2) \pm z_{.025}\sqrt{\hat{p}_1\hat{q}_1/n_1 + \hat{p}_2\hat{q}_2/n_2}$

   $.05 \pm 1.645\cdot\sqrt{(.07)(.93)/734 + (.02)(.98)/725}$

   $.05 \pm .02$

   $.03 < p_1-p_2 < .07$

   The interval does not contain zero, indicating a significant difference between the two dyspepsia rates.

2. Let those who saw someone else getting help be group 1.

   original claim: $p_1-p_2 > 0$

   $\hat{p}_1 = x_1/n_1 = 58/2000 = .0290$ $[x_1 = .0290\cdot2000 = 58]$

   $\hat{p}_2 = x_2/n_2 = 35/2000 = .0175$ $[x_2 = .0175\cdot2000 = 35]$

   $\hat{p}_1-\hat{p}_2 = .0290 - .0175 = .0115$

   $\bar{p} = (x_1 + x_2)/(n_1 + n_2)$

   $\quad = (58 + 35)/(2000 + 2000)$

   $\quad = 93/4000 = .02325$

   H_o: $p_1-p_2 \le 0$

   H_1: $p_1-p_2 > 0$

   $\alpha = .05$

   C.R. $z > z_{.05} = 1.645$

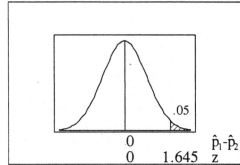

   calculations:

   $z_{\hat{p}1-\hat{p}2} = (\hat{p}_1-\hat{p}_2 - \mu_{\hat{p}1-\hat{p}2})/\sigma_{\hat{p}1-\hat{p}2}$

   $\quad = (.0115 - 0)/\sqrt{(.02325)(.97675)/2000 + (.02325)(.97675)/2000}$

   $\quad = .0115/.00477 = 2.413$

   conclusion:

   Reject H_o; there is sufficient evidence to conclude that $p_1-p_2 > 0$.

   P-value $= P(z > 2.41) = .5000 - .4920 = .0080$

3. Let the women be group 1. [n's > 30 and σ's known, use z]

   $\bar{x}_1 = 5$ hr 01 min $= 301$

   $\bar{x}_2 = 4$ hr 17 min $= 257$

   $\bar{x}_1-\bar{x}_2 = 301 - 257 = 44$

   $(\bar{x}_1-\bar{x}_2) \pm z_{.005}\sqrt{\sigma_1^2/n_1 + \sigma_2^2/n_2}$

   $44 \pm 2.575\sqrt{57^2/100 + 57^2/100}$

   44 ± 21

   $23 < \mu_1-\mu_2 < 65$

   The confidence interval does not contain 0, suggesting that there is a significant difference between the two means -- in fact, that the mean is greater for the women.

4. original claim: $\mu_d = 0$ [n $\le$ 30 and σ_d unknown, use t]

   $d = x_1 - x_2$: 5 0 0 0 8 1 1 4 0 1

   $n = 10$

   $\Sigma d = 20$ $\quad\quad\quad \bar{d} = 2.00$

   $\Sigma d^2 = 108$ $\quad\quad\quad s_d = 2.749$

   H_o: $\mu_d = 0$

   H_1: $\mu_d \ne 0$

   $\alpha = .05$ [assumed]

   C.R. $t < -t_{9,.025} = -2.262$

   $\quad\quad t > t_{9,.025} = 2.262$

   calculations:

   $t_{\bar{d}} = (\bar{d} - \mu_{\bar{d}})/s_{\bar{d}}$

   $\quad = (2.00 - 0)/(2.749/\sqrt{10})$

   $\quad = 2.00/.869 = 2.301$

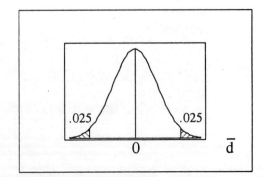

$-2.262 \quad 0 \quad 2.262 \quad t_9$

conclusion:
   Reject H_o; there is sufficient evidence to reject the claim that $\mu_d = 0$ and to conclude that $\mu_d \neq 0$ (in fact, that $\mu_d > 0$ -- i.e., that training reduces the weights).

5. Refer to the notation and data summary for exercise #4
   $\overline{d} \pm t_{9,.025} \cdot s_d/\sqrt{n}$
   $2.00 \pm 2.262 \cdot 2.749/\sqrt{10}$
   2.00 ± 1.97
   $0.0 < \mu_d < 4.0$

6. a. original claim: $\mu_1 - \mu_2 < 0$ [$n_1 > 30$ and $n_2 > 30$, use z (with s's for σ's)]
   $\overline{x}_1 - \overline{x}_2 = 26,588 - 44,765 = -18,177$
   H_o: $\mu_1 - \mu_2 \geq 0$
   H_1: $\mu_1 - \mu_2 < 0$
   $\alpha = .01$
   C.R. $z < -z_{.01} = -2.327$
   calculations:

   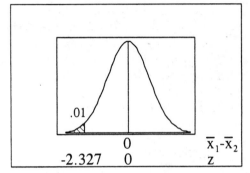

   $z_{\overline{x}_1 - \overline{x}_2} = (\overline{x}_1 - \overline{x}_2 - \mu_{\overline{x}_1 - \overline{x}_2})/\sigma_{\overline{x}_1 - \overline{x}_2}$
   $= (-18,177 - 0)/\sqrt{(8441)^2/85 + (12,469)^2/120}$
   $= -18,177/1460.78 = -12.443$

   conclusion:
      Reject H_o; there is sufficient evidence to conclude that $\mu_1 - \mu_2 < 0$.
      P-value $= P(z < -12.44) = .5000 - .4999 = .0001$

   b. $(\overline{x}_1 - \overline{x}_2) \pm z_{.005}\sqrt{\sigma_1^2/n_1 + \sigma_2^2/n_2}$
      $-18,177 \pm 2.575\sqrt{(8441)^2/100 + (12,469)^2/100}$
      $-18,177 \pm 3762$
      $-21,939 < \mu_1 - \mu_2 < -14,415$

7. a. original claim $\mu_d > 0$ [$n \leq 30$ and σ_d unknown, use t]
      $d = x_{after} - x_{before}$: 90 25 17 16 65 10 5 9 29 6 30 46
      $n = 12$
      $\Sigma d = 348$ $\overline{d} = 29.0$
      $\Sigma d^2 = 17594$ $s_d = 26.1$
      H_o: $\mu_d \leq 0$
      H_1: $\mu_d > 0$
      $\alpha = .05$
      C.R. $t > t_{11,.05} = 1.796$
      calculations:

      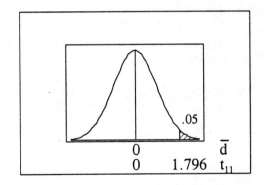

      $t_{\overline{d}} = (\overline{d} - \mu_{\overline{d}})/s_{\overline{d}}$
      $= (29.0 - 0)/(26.1/\sqrt{12})$
      $= 29.0/7.539 = 3.847$

      conclusion:
         Reject H_o; there is sufficient evidence to conclude that $\mu_d > 0$.
      b. $\overline{d} \pm t_{11,.025} \cdot s_{\overline{d}}$
         $29.0 \pm 2.201 \cdot 26.1/\sqrt{12}$
         29.0 ± 16.6
         $12.4 < \mu_d < 45.6$

Cumulative Review Exercises

1. Refer to the summary table at the right.

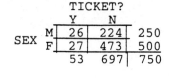

a. $P(Y) = 53/750 = .0707$
b. $P(M \text{ or } Y) = P(M) + P(Y) - P(M \text{ and } Y)$
$= 250/750 + 53/750 - 26/750$
$= 277/750 = .3693$
c. $P(Y \mid M) = 26/250 = .104$
d. $P(Y \mid F) = 27/500 = .054$
e. Let the males be group 1.
original claim: $p_1 - p_2 > 0$
$\hat{p}_1 = x_1/n_1 = 26/250 = .104$
$\hat{p}_2 = x_2/n_2 = 27/500 = .054$
$\hat{p}_1 - \hat{p}_2 = .104 - .054 = .050$
$\bar{p} = (x_1 + x_2)/(n_1 + n_2)$
$= (26 + 27)/(250 + 500)$
$= 53/750 = .0707$
$H_o: p_1 - p_2 \le 0$
$H_1: p_1 - p_2 > 0$
$\alpha = .05$
C.R. $z > z_{.05} = 1.645$

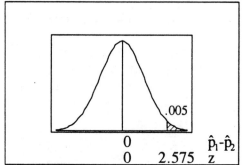

calculations:
$z_{\hat{p}_1 - \hat{p}_2} = (\hat{p}_1 - \hat{p}_2 - \mu_{\hat{p}_1 - \hat{p}_2})/\sigma_{\hat{p}_1 - \hat{p}_2}$
$= (.050 - 0)/\sqrt{(.0707)(.9293)/250 + (.0707)(.9293)/500}$
$= .050/.0199 = 2.519$
conclusion:
   Reject H_o; there is sufficient evidence to conclude that $p_1 - p_2 > 0$.
P-value $= P(z > 2.52) = .5000 - .4941 = .0059$
No; we cannot conclude that men speed more, only that they are ticketed more. It is possible, for example, that men drive more.

2. There is a problem with the reported results, and no statistical analysis would be appropriate. Since there were 100 drivers in each group, and the number in each group owning a cell phone must be a whole number between 0 and 100 inclusive, the sample proportion for each group must be a whole percent. The reported values of 13.7% and 10.6% are not mathematical possibilities for the sample success rates of groups of 100.

3. Let the prepared students be group 1.

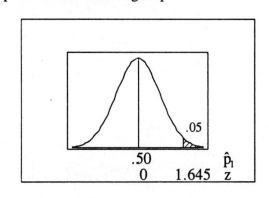

a. original claim: $p_1 > .50$
$\hat{p}_1 = x_1/n_1 = 62/80 = .775$
$H_o: p_1 \le .50$
$H_1: p_1 > .50$
$\alpha = .05$
C.R. $z > z_{.05} = 1.645$
calculations:
$z_{\hat{p}_1} = (\hat{p}_1 - \mu_{\hat{p}_1})/\sigma_{\hat{p}_1}$
$= (.775 - .50)/\sqrt{(.50)(.50)/80}$
$= .275/.0559 = 4.919$
conclusion:
   Reject H_o; there is sufficient evidence to conclude that $p_1 > .50$.

b. original claim: p_1-p_2 > 0 [concerns p's, use z]

$\hat{p}_1 = x_1/n_1 = 62/80 = .775$

$\hat{p}_2 = x_2/n_2 = 23/50 = .460$

$\hat{p}_1$-$\hat{p}_2 = .775 - .460 = .315$

$\bar{p} = (x_1 + x_2)/(n_1 + n_2)$

$= (62 + 23)/(80 + 50)$

$= 85/130 = .654$

H_0: p_1-$p_2 \leq 0$

H_1: p_1-$p_2 > 0$

$\alpha = .05$

C.R. z > $z_{.05}$ = 1.645

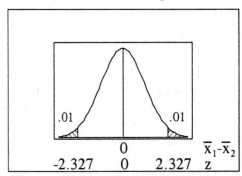

calculations:

$z_{\hat{p}_1-\hat{p}_2} = (\hat{p}_1$-$\hat{p}_2 - \mu_{\hat{p}_1-\hat{p}_2})/\sigma_{\hat{p}_1-\hat{p}_2}$

$= (.315 - 0)/\sqrt{(.654)(.346)/80 + (.654)(.346)/50}$

$= .315/.0858 = 3.673$

conclusion:

Reject H_0; there is sufficient evidence to conclude that p_1-p_2 > 0.

4. Let the women be group 1.

a. original claim: μ_1-$\mu_2 = 0$ [n_1 > 30 and n_2 > 30, use z (with s's for σ's)]

$\bar{x}_1$-$\bar{x}_2 = 538.82 - 525.23 = 13.59$

H_0: μ_1-$\mu_2 = 0$

H_1: μ_1-$\mu_2 \neq 0$

$\alpha = .02$

C.R. z < -$z_{.01}$ = -2.327

z > $z_{.01}$ = 2.327

calculations:

$z_{\bar{x}_1-\bar{x}_2} = (\bar{x}_1$-$\bar{x}_2 - \mu_{\bar{x}_1-\bar{x}_2})/\sigma_{\bar{x}_1-\bar{x}_2}$

$= (13.59 - 0)/\sqrt{(114.16)^2/68 + (97.23)^2/86}$

$= 13.59/17.366 = .783$

conclusion:

Do not reject H_0; there is not sufficient evidence to reject the claim that μ_1-$\mu_2 = 0$.

P-value = $2 \cdot P(z > .78) = 2 \cdot (.5000 - .2823) = 2 \cdot (.2177) = .4354$

b. $\bar{x}_1 \pm z_{.01} \cdot \sigma_{\bar{x}_1}$

$538.82 \pm 2.327 \cdot 114.16/\sqrt{68}$

538.82 ± 32.21

$506.61 < \mu < 571.03$

Chapter 9

Correlation and Regression

9-2 Correlation

1. a. From Table A-6, CV = ±.312; therefore r = .401 indicates a significant (positive) linear correlation.
   b. The proportion of the variation in y that can be explained in terms of the variation in x is r_2 = $(.401)^2$ = .1608, or 16.08%.

NOTE: In addition to the value of n, calculation of r requires five sums: Σx, Σy, Σx^2, Σy^2 and Σxy. The next problem shows the chart prepared to find these sums. As the sums can usually be found conveniently using a calculator and without constructing the chart, subsequent problems typically give only the values of the sums and do not show a chart.

Also, calculation of r also involves three subcalculations.
   (1) $n(\Sigma xy) - (\Sigma x)(\Sigma y)$ determines the sign of r. If large values of x are associated with large values of y, it will be positive. If large values of x are associated with small values of y, it will be negative. If not, a mistake has been made.
   (2) $n(\Sigma x^2) - (\Sigma x)^2$ cannot be negative. If it is, a mistake has been made.
   (3) $n(\Sigma y^2) - (\Sigma y)^2$ cannot be negative. If it is, a mistake has been made.

Finally, r must be between -1 and 1 inclusive. If not, a mistake has been made. If this or any of the previous mistakes occurs, stop immediately and find the error; continuing will be a fruitless waste of effort.

3.

x	y	xy	x^2	y^2
1	19	19	1	361
1	12	24	4	144
3	4	12	9	16
5	18	90	25	324
5	20	100	25	400
16	73	245	64	1245

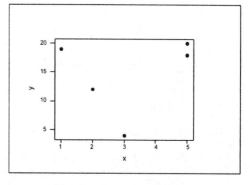

$n(\Sigma xy) - (\Sigma x)(\Sigma y) = 5(245) - (16)(73) = 57$
$n(\Sigma x^2) - (\Sigma x)^2 = 5(64) - (16)^2 = 64$
$n(\Sigma y^2) - (\Sigma y)^2 = 5(1245) - (73)^2 = 896$

   a. According to the scatter diagram, there appears to be a significant "U-shaped" pattern, but no <u>linear</u> relationship between x and y. Expect a value for r close to 0.
   b. $r = [n(\Sigma xy) - (\Sigma x)(\Sigma y)]/[\sqrt{n(\Sigma x^2) - (\Sigma x)^2} \cdot \sqrt{n(\Sigma y^2) - (\Sigma y)^2}]$
      $= 57/[\sqrt{64} \cdot \sqrt{896}]$
      $= .238$
   From Table A-6, CV = ±.878; therefore r = .238 does not indicates a significant linear correlation. This agrees with the interpretation of the scatter diagram in part (a).

5. a. n = 9
$\Sigma x = 632.0$
$\Sigma y = 1089$
$\Sigma x^2 = 44399.50$
$\Sigma y^2 = 132223$
$\Sigma xy = 76546.0$

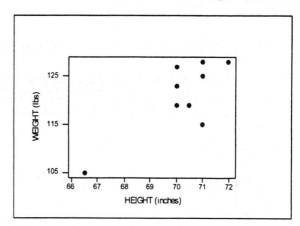

b. $n(\Sigma xy) - (\Sigma x)(\Sigma y) = 9(76546) - (632)(1089) = 666$
$n(\Sigma x^2) - (\Sigma x)^2 = 9(44399.5) - (632)^2 = 171.5$
$n(\Sigma y^2) - (\Sigma y)^2 = 9(132223) - (1089)^2 = 4086$
$r = [n(\Sigma xy) - (\Sigma x)(\Sigma y)]/[\sqrt{n(\Sigma x^2) - (\Sigma x)^2} \cdot \sqrt{n(\Sigma y^2) - (\Sigma y)^2}]$
$= 666/[\sqrt{171.5} \cdot \sqrt{4086}]$
$= .796$

c. $H_o: \rho = 0$
$H_1: \rho \neq 0$
$\alpha = .05$
C.R. $r < -.666$ OR C.R. $t < -t_{7,.025} = -2.365$
$r > .666$ $t > t_{7,.025} = 2.365$

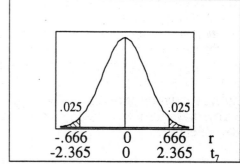

calculations: calculations:
$r = .796$ $t_r = (r - \mu_r)/s_r$
$= (.796 - 0)/\sqrt{(1-(.796)^2)/7}$
$= .796/.229 = 3.475$

conclusion:
Reject H_o; there is sufficient evidence to reject the claim that $\rho = 0$ and to conclude that $\rho \neq 0$ (in fact, that $\rho > 0$).

7. a. n = 8
$\Sigma x = 14.60$
$\Sigma y = 26$
$\Sigma x^2 = 32.9632$
$\Sigma y^2 = 104$
$\Sigma xy = 56.80$

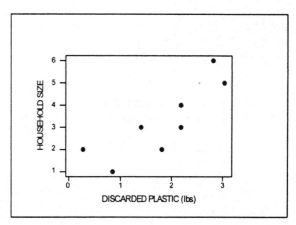

b. $n(\Sigma xy) - (\Sigma x)(\Sigma y) = 8(56.80) - (14.60)(26) = 74.80$
$n(\Sigma x^2) - (\Sigma x)^2 = 8(32.9632) - (14.60)^2 = 50.5456$
$n(\Sigma y^2) - (\Sigma y)^2 = 8(104) - (26)^2 = 156$
$r = [n(\Sigma xy) - (\Sigma x)(\Sigma y)]/[\sqrt{n(\Sigma x^2) - (\Sigma x)^2} \cdot \sqrt{n(\Sigma y^2) - (\Sigma y)^2}]$
$= 74.80/[\sqrt{50.5456} \cdot \sqrt{156}]$
$= .842$

c. $H_o: \rho = 0$
$H_1: \rho \neq 0$
$\alpha = .05$
C.R. $r < -.707$ $\quad$ <u>OR</u> $\quad$ C.R. $t < -t_{6,.025} = -2.447$
$\quad\quad r > .707$ $\quad\quad\quad\quad\quad\quad\quad t > t_{6,.025} = 2.447$

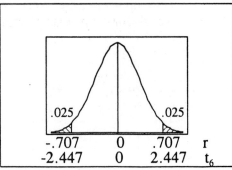

calculations:
$\quad r = .842$

calculations:
$\quad t_r = (r - \mu_r)/s_r$
$\quad\quad = (.842 - 0)/\sqrt{(1-(.842)^2)/6}$
$\quad\quad = .842/.220 = 3.829$

conclusion:
$\quad$ Reject H_o; there is sufficient evidence to reject the claim that $\rho = 0$ and to conclude that $\rho \neq 0$ (in fact, $\rho > 0$).

9. a. $n = 8$
$\Sigma x = 276.6$
$\Sigma y = 1.66$
$\Sigma x^2 = 10680.48$
$\Sigma y^2 = .3522$
$\Sigma xy = 57.191$

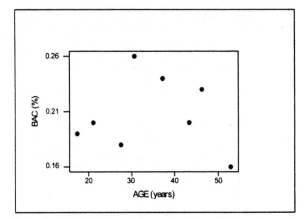

b. $n(\Sigma xy) - (\Sigma x)(\Sigma y) = 8(57.191) - (276.6)(1.66) = -1.628$
$n(\Sigma x^2) - (\Sigma x)^2 = 8(10680.48) - (276.6)^2 = 8936.28$
$n(\Sigma y^2) - (\Sigma y)^2 = 8(.3522) - (1.66)^2 = .062$
$r = [n(\Sigma xy) - (\Sigma x)(\Sigma y)]/[\sqrt{n(\Sigma x^2) - (\Sigma x)^2} \cdot \sqrt{n(\Sigma y^2) - (\Sigma y)^2}\,]$
$\quad = -1.628/[\sqrt{8936.28} \cdot \sqrt{.062}\,]$
$\quad = -.069$

c. $H_o: \rho = 0$
$H_1: \rho \neq 0$
$\alpha = .05$
C.R. $r < -.707$ $\quad$ <u>OR</u> $\quad$ C.R. $t < -t_{6,.025} = -2.447$
$\quad\quad r > .707$ $\quad\quad\quad\quad\quad\quad\quad t > t_{6,.025} = 2.447$

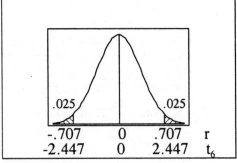

calculations:
$\quad r = -.069$

calculations:
$\quad t_r = (r - \mu_r)/s_r$
$\quad\quad = (-.069 - 0)/\sqrt{(1-(-.069)^2)/6}$
$\quad\quad = -.069/.407 = -.170$

conclusion:
$\quad$ Do not reject H_o; there is not sufficient evidence to reject the claim that $\rho = 0$.
No; based on this result, the BAC level does not seem to be related to the age of the person tested.

11. A. Interval-duration relationship?
   a. $n = 50$
   $\Sigma x = 10832$
   $\Sigma y = 4033$
   $\Sigma x^2 = 2513280$
   $\Sigma y^2 = 332331$
   $\Sigma xy = 903488$

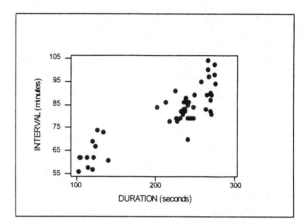

b. $n(\Sigma xy) - (\Sigma x)(\Sigma y) = 50(903488) - (10832)(4033) = 1488944$
   $n(\Sigma x^2) - (\Sigma x)^2 = 50(2513280) - (10832)^2 = 8331776$
   $n(\Sigma y^2) - (\Sigma y)^2 = 50(332331) - (4033)^2 = 351461$
   $r = [n(\Sigma xy) - (\Sigma x)(\Sigma y)]/[\sqrt{n(\Sigma x^2) - (\Sigma x)^2} \cdot \sqrt{n(\Sigma y^2) - (\Sigma y)^2}]$
   $= 1488944/[\sqrt{8331776} \cdot \sqrt{351461}]$
   $= .870$

c. $H_o: \rho = 0$
   $H_1: \rho \neq 0$
   $\alpha = .05$
   C.R. $r < -.279$ OR   C.R. $t < -t_{48,.025} = -1.960$
   $ r > .279$ $ t > t_{48,.025} = 1.960$

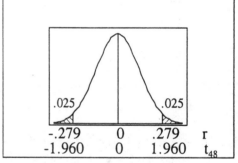

calculations: calculations:
   $r = .870$ $t_r = (r - \mu_r)/s_r$
    $= (.870 - 0)/\sqrt{(1-(.870)^2)/48}$
    $= .870/.0711 = 12.231$

conclusion:
   Reject H_o; there is sufficient evidence to reject the claim that $\rho = 0$ and to conclude that $\rho \neq 0$ (in fact, that $\rho > 0$).
   Yes; there is a significant positive linear correlation, suggesting that the interval after an eruption is related to the duration of the eruption. NOTE: The longer the duration of an eruption, the more pressure has been released and the longer it will take the geyser to build back up for another eruption. In fact, the park rangers use the duration of one eruption to predict the time of the next eruption.

B. Interval-height relationship?
   a. $n = 50$
   $\Sigma x = 6904$
   $\Sigma y = 4033$
   $\Sigma x^2 = 961150$
   $\Sigma y^2 = 332331$
   $\Sigma xy = 556804$

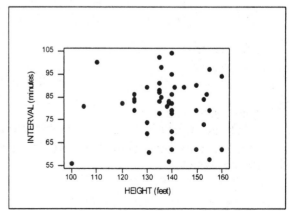

b. $n(\Sigma xy) - (\Sigma x)(\Sigma y) = 50(556804) - (6904)(4033) = -3632$
$n(\Sigma x^2) - (\Sigma x)^2 = 50(961150) - (6904)^2 = 392284$
$n(\Sigma y^2) - (\Sigma y)^2 = 50(332331) - (4033)^2 = 351461$
$r = [n(\Sigma xy) - (\Sigma x)(\Sigma y)]/[\sqrt{n(\Sigma x^2) - (\Sigma x)^2} \cdot \sqrt{n(\Sigma y^2) - (\Sigma y)^2}]$
$= -3632/[\sqrt{392284} \cdot \sqrt{351461}]$
$= -.00978$

c. $H_o: \rho = 0$
$H_1: \rho \neq 0$
$\alpha = .05$
C.R. $r < -.279$ <u>OR</u>   C.R. $t < -t_{48,.025} = -1.960$
$r > .279$ $t > t_{48,.025} = 1.960$

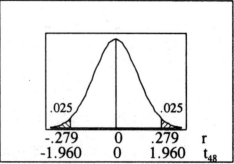

calculations: calculations:
$r = -.00978$ $t_r = (r - \mu_r)/s_r$
$= (-.00978 - 0)/\sqrt{(1-(-.00978)^2)/48}$
$= -.00978/.144 = -.068$

conclusion:
Do not reject H_o; there is not sufficient evidence to reject the claim that $\rho = 0$.
No; there is not a significant linear correlation, suggesting that the interval after an eruption is not so related to the height of the eruption.

C. Duration is the more relevant predictor of the interval until the next eruption -- because it has a significant correlation with interval, while height does not.

13. A. Price-weight relationship? (NOTE: For convenience, we measure the price y in $1000's)

a. $n = 30$
$\Sigma x = 50.65$
$\Sigma y = 433.23$
$\Sigma x^2 = 109.74$
$\Sigma y^2 = 12378.86$
$\Sigma xy = 1027.03$

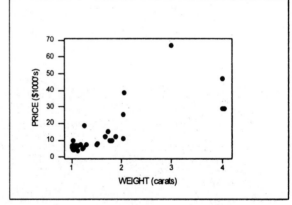

b. $n(\Sigma xy) - (\Sigma x)(\Sigma y) = 30(1027.03) - (50.65)(433.23) = 8867.80$
$n(\Sigma x^2) - (\Sigma x)^2 = 30(109.74) - (50.65)^2 = 726.78$
$n(\Sigma y^2) - (\Sigma y)^2 = 30(12378.86) - (433.23)^2 = 183677.57$
$r = [n(\Sigma xy) - (\Sigma x)(\Sigma y)]/[\sqrt{n(\Sigma x^2) - (\Sigma x)^2} \cdot \sqrt{n(\Sigma y^2) - (\Sigma y)^2}]$
$= 8867.80/[\sqrt{726.78} \cdot \sqrt{183677.57}]$
$= .767$

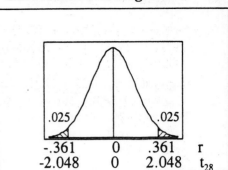

c. $H_o: \rho = 0$
   $H_1: \rho \neq 0$
   $\alpha = .05$
   C.R. $r < -.361$ **OR**   C.R. $t < -t_{28,.025} = -2.048$
   $\quad\; r > .361$ $\qquad\qquad t > t_{28,.025} = 2.048$

calculations: calculations:
   $r = .767$ $\qquad\qquad t_r = (r - \mu_r)/s_r$
   $\qquad\qquad\qquad\quad = (.767 - 0)/\sqrt{(1-(.767)^2)/28}$
   $\qquad\qquad\qquad\quad = .767/.121 = 6.325$

conclusion:
   Reject H_o; there is sufficient evidence to reject the claim that $\rho = 0$ and to conclude
   that $\rho \neq 0$ (in fact, that $\rho > 0$).

Yes; based on this result there is a significant positive linear correlation between the
price of a diamond and its weight in carats.

B. Price-color relationship? (NOTE: For convenience, we measure the price y in $1000's)

a. n = 30
   Σx = 140
   Σy = 433.23
   Σx² = 782
   Σy² = 12378.86
   Σxy = 1630.03

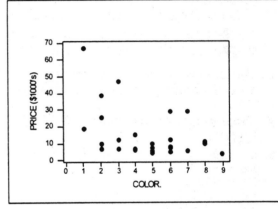

b. $n(\Sigma xy) - (\Sigma x)(\Sigma y) = 30(1630.03) - (140)(433.23) = -11751.30$
   $n(\Sigma x^2) - (\Sigma x)^2 = 30(782) - (140)^2 = 3860$
   $n(\Sigma y^2) - (\Sigma y)^2 = 30(12378.86) - (433.23)^2 = 183677.57$
   $r = [n(\Sigma xy) - (\Sigma x)(\Sigma y)]/[\sqrt{n(\Sigma x^2) - (\Sigma x)^2} \cdot \sqrt{n(\Sigma y^2) - (\Sigma y)^2}\,]$
   $\quad = -11751.30/[\sqrt{3860} \cdot \sqrt{183677.57}\,]$
   $\quad = -.441$

c. $H_o: \rho = 0$
   $H_1: \rho \neq 0$
   $\alpha = .05$
   C.R. $r < -.361$ **OR**   C.R. $t < -t_{28,.025} = -2.048$
   $\quad\; r > .361$ $\qquad\qquad t > t_{28,.025} = 2.048$

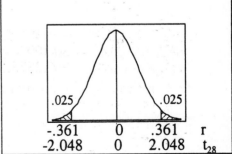

calculations: calculations:
   $r = -.441$ $\qquad\qquad t_r = (r - \mu_r)/s_r$
   $\qquad\qquad\qquad\quad = (-.441 - 0)/\sqrt{(1-(-.441)^2)/28}$
   $\qquad\qquad\qquad\quad = -.441/.170 = -2.600$

conclusion:
   Reject H_o; there is sufficient evidence to reject the claim that $\rho = 0$ and to conclude
   that $\rho \neq 0$ (in fact, that $\rho < 0$).

Yes; based on this result there is a significant negative linear correlation between the price of a diamond and its color -- i.e., the lower the color number, the higher the price.

C. When determining the value of a diamond, weight is more important than color. The weight of a diamond is more highly correlated (i.e., has an r value farther from 0) with the selling price than is the color.

15. A. Gross-budget relationship?
   a. n = 36
      Σx = 1931.375
      Σy = 4966.978
      Σx^2 = 173580.583
      Σy^2 = 1153595.699
      Σxy = 338719.616

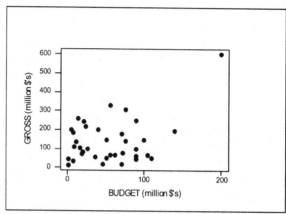

   b. $n(\Sigma xy) - (\Sigma x)(\Sigma y) = 36(338719.616) - (1931.375)(4966.978) = 2600808.825$
      $n(\Sigma x^2) - (\Sigma x)^2 = 36(173580.583) - (1931.375)^2 = 2518691.489$
      $n(\Sigma y^2) - (\Sigma y)^2 = 36(1153595.699) - (4966.978)^2 = 16858571.148$
      $r = [n(\Sigma xy) - (\Sigma x)(\Sigma y)]/[\sqrt{n(\Sigma x^2) - (\Sigma x)^2} \cdot \sqrt{n(\Sigma y^2) - (\Sigma y)^2}]$
       $= 2600808.825/[\sqrt{2518691.489} \cdot \sqrt{16858571.148}]$
       $= .399$

   c. $H_o: \rho = 0$
      $H_1: \rho \neq 0$
      $\alpha = .05$
      C.R. r < -.335   OR   C.R. t < $-t_{34,.025}$ = -1.960
         r > .335   t > $t_{34,.025}$ = 1.960

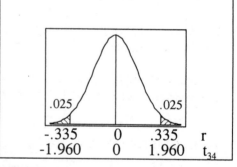

.025   .025

-.335   0   .335   r
-1.960   0   1.960 t_{34}

   calculations: calculations:
      r = .399 $t_r = (r - \mu_r)/s_r$
                            $= (.399 - 0)/\sqrt{(1-(.399)^2)/34}$
                            $= .399/.157 = 2.538$
   conclusion:
      Reject H_o; there is sufficient evidence to reject the claim that $\rho = 0$ and to conclude that $\rho \neq 0$ (in fact, that $\rho > 0$).

B. Gross-rating relationship?
   a. n = 36
      Σx = 267.2
      Σy = 4966.978
      Σx^2 = 2017.40
      Σy^2 = 1153595.699
      Σxy = 37995.901

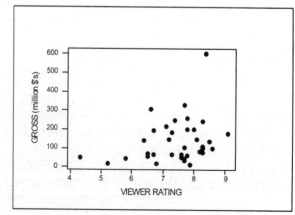

b. $n(\Sigma xy) - (\Sigma x)(\Sigma y) = 36(37995.901) - (267.2)(4966.978) = 40675.914$
   $n(\Sigma x^2) - (\Sigma x)^2 = 36(2017.40) - (267.2)^2 = 1230.56$
   $n(\Sigma y^2) - (\Sigma y)^2 = 36(1153595.699) - (4966.978)^2 = 16858571.148$
   $r = [n(\Sigma xy) - (\Sigma x)(\Sigma y)]/[\sqrt{n(\Sigma x^2) - (\Sigma x)^2} \cdot \sqrt{n(\Sigma y^2) - (\Sigma y)^2}]$
     $= 40675.914/[\sqrt{1230.56} \cdot \sqrt{16858571.148}]$
     $= .282$

c. $H_o: \rho = 0$
   $H_1: \rho \neq 0$
   $\alpha = .05$
   C.R. $r < -.335$ OR   C.R. $t < -t_{34,.025} = -1.960$
      $r > .335$ $t > t_{34,.025} = 1.960$

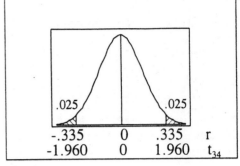

calculations: calculations:
   $r = .282$ $t_r = (r - \mu_r)/s_r$
                              $= (.282 - 0)/\sqrt{(1-(.282)^2)/34}$
                              $= .282/.165 = 1.717$

conclusion:
   Do not reject H_o; there is not sufficient evidence to reject the claim that $\rho = 0$.
   No, there is not sufficient evidence to say there is a significant linear correlation between these two variables.

C. Based on the preceding results, invest in the big budget movies. There is a stronger correlation between budget and gross than between viewer rating and gross. It appears that the publicity associated with big budgets attracts more patrons than do high ratings by other viewers.

17. A linear correlation coefficient very close to zero indicates <u>no</u> significant linear correlation and no tendencies can be inferred.

19. A linear correlation coefficient very close to zero indicates no significant <u>linear</u> correlation, but there may some other type of relationship between the variables.

21. There are n=13 data points.
    From the scatterplot, it appears that the correlation is about +.75.
    $H_o: \rho = 0$
    $H_1: \rho \neq 0$
    $\alpha = .05$
    C.R. $r < -.553$ OR   C.R. $t < -t_{11,.025} = -2.201$
       $r > .553$ $t > t_{11,.025} = 2.201$

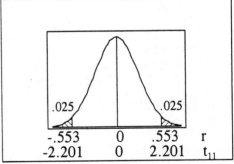

calculations: calculations:
   $r = .75$ $t_r = (r - \mu_r)/s_r$
                              $= (.75 - 0)/\sqrt{(1-(.75)^2)/11}$
                              $= .75/.199 = 3.761$

conclusion:
   Reject H_o; there is sufficient evidence to reject the claim that $\rho = 0$ and to conclude that $\rho \neq 0$ (in fact, that $\rho > 0$).
   Yes; based on this result there is a significant positive linear correlation between the mortality rate and the rate of infections acquired in intensive care units.

23. a. For $\pm t_{48,.025} = \pm 1.960$,
    the critical values are $r = \pm 1.960/\sqrt{(\pm 1.960)^2 + 48} = \pm .272$.

    b. For $\pm t_{73,.05} = \pm 1.645$,
    the critical values are $r = \pm 1.645/\sqrt{(\pm 1.645)^2 + 73} = \pm .189$.

    c. For $-t_{18,.05} = -1.734$,
    the critical value is $r = -1.734/\sqrt{(-1.734)^2 + 18} = -.378$.

    d. For $t_{8,.05} = 1.860$,
    the critical value is $r = 1.860/\sqrt{(1.860)^2 + 8} = .549$.

    e. For $t_{10,.01} = 2.764$,
    the critical value is $r = 2.764/\sqrt{(2.764)^2 + 10} = .658$.

25. No; correlation cannot be used to address the issue of whether male statistics students exaggerate their heights. The two concepts are not related. A high positive correlation near $r = 1.000$, for example, indicates either that the reported and actual heights tended to agree or that the reported heights were over-stated (or under-stated) by a consistent linear pattern (e.g., each student over-stating by approximately 1.5").
    NOTE: The underlying principle is that a high correlation indicates that one value can be predicted from the other -- not that the values are equal. A related and often incorrectly analyzed scenario is that a high correlation between grades on test 1 and test 2 does not mean that each student received approximately the same grade on the two tests -- only that there is some linear relationship that allows us to predict one score if we know the other.

9-3 Regression

NOTE: For exercises 1-20, the exact summary statistics (i.e., without any rounding) are given on the right. While the intermediate calculations on the left are presented rounded to various degrees of accuracy, the entire unrounded values were preserved in the calculator until the very end.

1. $\bar{x} = 3.00$ $n = 4$
   $\bar{y} = 8.00$ $\Sigma x = 12$
   $b_1 = [n(\Sigma xy) - (\Sigma x)(\Sigma y)]/[n(\Sigma x^2) - (\Sigma x)^2]$ $\Sigma y = 32$
   $\quad = [4(104) - (12)(32)]/[4(44) - (12)^2]$ $\Sigma x^2 = 44$
   $\quad = 32/32$ $\Sigma y^2 = 264$
   $\quad = 1.00$ $\Sigma xy = 104$
   $b_0 = \bar{y} - b_1\bar{x}$
   $\quad = 8.00 - (1.00)(3.00)$
   $\quad = 5.00$
   $\hat{y} = b_0 + b_1 x = 5.00 + 1.00x$

3. $\bar{x} = 3.20$ $n = 5$
   $\bar{y} = 14.60$ $\Sigma x = 16$
   $b_1 = [n(\Sigma xy) - (\Sigma x)(\Sigma y)]/[n(\Sigma x^2) - (\Sigma x)^2]$ $\Sigma y = 73$
   $\quad = [5(245) - (16)(73)]/[5(64) - (16)^2]$ $\Sigma x^2 = 64$
   $\quad = 57/64$ $\Sigma y^2 = 1245$
   $\quad = 0.89$ $\Sigma xy = 245$
   $b_0 = \bar{y} - b_1\bar{x}$
   $\quad = 14.60 - (0.89)(3.20)$
   $\quad = 11.75$
   $\hat{y} = b_0 + b_1 x = 11.75 + 0.89x$

NOTE: In the exercises that follow, this manual uses the full accuracy of b_0 and b_1 when calculating $\hat{y}$. Using only the rounded values as stated in the equation produces slightly different answers.

5. $\bar{x} = 70.22$
   $\bar{y} = 121.00$
   $b_1 = [n(\Sigma xy) - (\Sigma x)(\Sigma y)]/[n(\Sigma x^2) - (\Sigma x)^2]$
   $= 666.0/171.50$
   $= 3.88$
   $b_0 = \bar{y} - b_1\bar{x}$
   $= 121.00 - (3.88)(70.22)$
   $= -151.70$
   $\hat{y} = b_0 + b_1 x = -151.7 + 3.88x$
   $\hat{y}_{69} = -151.7 + 3.88(69) = 116.3$ lbs

   $n = 9$
   $\Sigma x = 632.0$
   $\Sigma y = 1089$
   $\Sigma x^2 = 44399.50$
   $\Sigma y^2 = 132223$
   $\Sigma xy = 76546.0$

7. $\bar{x} = 1.825$
   $\bar{y} = 3.25$
   $b_1 = [n(\Sigma xy) - (\Sigma x)(\Sigma y)]/[n(\Sigma x^2) - (\Sigma x)^2]$
   $= 74.80/50.5456$
   $= 1.480$
   $b_0 = \bar{y} - b_1\bar{x}$
   $= 3.25 - (1.480)(1.825)$
   $= .549$
   $\hat{y} = b_0 + b_1 x = .549 + 1.480x$
   $\hat{y}_{.50} = .549 + 1.480(.50) = 1.3$ people

   $n = 8$
   $\Sigma x = 14.60$
   $\Sigma y = 26$
   $\Sigma x^2 = 32.9632$
   $\Sigma y^2 = 104$
   $\Sigma xy = 56.80$

9. $\bar{x} = 34.575$
   $\bar{y} = .2075$
   $b_1 = [n(\Sigma xy) - (\Sigma x)(\Sigma y)]/[n(\Sigma x^2) - (\Sigma x)^2]$
   $= -1.628/8936.28$
   $= -.000182$
   $b_0 = \bar{y} - b_1\bar{x}$
   $= .2075 - (-.000182)(34.575)$
   $= .214$
   $\hat{y} = b_0 + b_1 x = .214 - .000182x$
   $\hat{y}_{21} = \bar{y} = .2075\%$ [no significant correlation]

   $n = 8$
   $\Sigma x = 276.6$
   $\Sigma y = 1.66$
   $\Sigma x^2 = 10680.48$
   $\Sigma y^2 = .3522$
   $\Sigma xy = 57.191$

11. a. interval/duration relationship
    $\bar{x} = 216.64$
    $\bar{y} = 80.66$
    $b_1 = [n(\Sigma xy) - (\Sigma x)(\Sigma y)]/[n(\Sigma x^2) - (\Sigma x)^2]$
    $= 1488944/8331776$
    $= .179$
    $b_0 = \bar{y} - b_1\bar{x}$
    $= 80.66 - (.179)(216.64)$
    $= 41.9$
    $\hat{y} = b_0 + b_1 x = 41.9 + .179x$
    $\hat{y}_{210} = 41.9 + .179(210) = 79.5$ minutes

    $n = 50$
    $\Sigma x = 10832$
    $\Sigma y = 4033$
    $\Sigma x^2 = 2513280$
    $\Sigma y^2 = 332331$
    $\Sigma xy = 903488$

b. interval/height relationship

$\bar{x} = 138.08$ $n = 50$
$\bar{y} = 80.66$ $\Sigma x = 6904$
$b_1 = [n(\Sigma xy) - (\Sigma x)(\Sigma y)]/[n(\Sigma x^2) - (\Sigma x)^2]$ $\Sigma y = 4033$
$\quad = -3632/392284$ $\Sigma x^2 = 961150$
$\quad = -.00926$ $\Sigma y^2 = 332331$
$b_0 = \bar{y} - b_1\bar{x}$ $\Sigma xy = 556804$
$\quad = 80.66 - (-.00926)(138.08)$
$\quad = 81.9$
$\hat{y} = b_0 + b_1x = 81.9 - .00926x$
$\hat{y}_{275} = \bar{y} = 80.7$ minutes [no significant correlation]

c. The predicted time in part (a) is better, since interval and duration are significantly correlated. Since interval and height are not significantly correlated, the predicted time in part (b) did not even use the height data.

13. a. price/weight relationship [NOTE: For convenience, we measure the price y in $1000's.)

$\bar{x} = 1.688$ $n = 30$
$\bar{y} = 14.441$ $\Sigma x = 50.65$
$b_1 = [n(\Sigma xy) - (\Sigma x)(\Sigma y)]/[n(\Sigma x^2) - (\Sigma x)^2]$ $\Sigma y = 433.23$
$\quad = 8867.80/726.78$ $\Sigma x^2 = 109.74$
$\quad = 12.202$ $\Sigma y^2 = 12378.86$
$b_0 = \bar{y} - b_1\bar{x}$ $\Sigma xy = 1027.03$
$\quad = 14.441 - 12.202(1.688)$
$\quad = -6.159$
$\hat{y} = b_0 + b_1x = -6.159 + 12.202x$
$\hat{y}_{1.5} = -6.159 + 12.202(1.5) = 12.143$ [$12,143]

b. price/color relationship (NOTE: For convenience, we measure the price y in $100's.)

$\bar{x} = 4.667$ $n = 50$
$\bar{y} = 14.441$ $\Sigma x = 140$
$b_1 = [n(\Sigma xy) - (\Sigma x)(\Sigma y)]/[n(\Sigma x^2) - (\Sigma x)^2]$ $\Sigma y = 433.23$
$\quad = -11751.3/3860$ $\Sigma x^2 = 782$
$\quad = -3.044$ $\Sigma y^2 = 12378.86$
$b_0 = \bar{y} - b_1\bar{x}$ $\Sigma xy = 1630.03$
$\quad = 14.441 - (-3.044)(4.667)$
$\quad = 28.648$
$\hat{y} = b_0 + b_1x = 28.648 + (-3.044)x$
$\hat{y}_3 = 28.648 - 3.044(3) = 19.515$ [$19,515]

c. The predicted time in part (a) is better, since there is a stronger correlation between price and weight than between price and color.

15. a. gross/budget relationship

$\bar{x} = 53.649$ $n = 36$
$\bar{y} = 137.972$ $\Sigma x = 1931.375$
$b_1 = [n(\Sigma xy) - (\Sigma x)(\Sigma y)]/[n(\Sigma x^2) - (\Sigma x)^2]$ $\Sigma y = 4966.978$
$\quad = 2600808.8/2518691.5$ $\Sigma x^2 = 173580.583$
$\quad = 1.033$ $\Sigma y^2 = 1153595.699$
$b_0 = \bar{y} - b_1\bar{x}$ $\Sigma xy = 338719.616$
$\quad = 137.972 - 1.033(53.649)$
$\quad = 82.573$
$\hat{y} = b_0 + b_1x = 82.573 + 1.033x$
$\hat{y}_{15} = 82.573 + 1.033(15) = 98.06$ [$98,060,000]

b. gross/rating relationship

$\bar{x} = 7.422$

$\bar{y} = 137.972$

$b_1 = [n(\Sigma xy) - (\Sigma x)(\Sigma y)]/[n(\Sigma x^2) - (\Sigma x)^2]$

    $= 40675.91/1230.56$

    $= 33.055$

$b_0 = \bar{y} - b_1\bar{x}$

    $= 137.972 - 33.055(7.422)$

    $= -107.368$

$\hat{y} = b_0 + b_1x = -107.368 + 33.055x$

$\hat{y}_7 = \bar{y} = 137.97$ [$137,970,000] [no signiificant correlation]

$n = 30$

$\Sigma x = 267.2$

$\Sigma y = 4966.978$

$\Sigma x^2 = 2017.40$

$\Sigma y^2 = 1153595.699$

$\Sigma xy = 37995.901$

c. The predicted time in part (a) is better, since gross and budget are significantly correlated. Since gross and rating are not significantly correlated, the predicted gross in part (b) did not even use the rating data.

17. The .05 critical values for r are taken from Table A-6.
   a. CV $= \pm.632$; r $= .931$ is significant
   use $\hat{y} = 4.00 + 2.00x$
       $\hat{y}_{3.00} = 4.00 + 2.00(3.00) = 10.00$
   b. CV $= \pm.220$; r $= -.033$ is not significant
   use $\hat{y} = \bar{y}$
       $\hat{y}_{3.00} = \bar{y} = 2.50$

19. a. Yes; the point is an outlier. It is far away from the other data points; $400 is considerably larger than the other bills, and $50 is considerably larger than the other tips.
   b. No; the point is not an influential one. The original regression line $\hat{y} = -.347 + .149x$ predicts $\hat{y}_{400} = -.347 + .149(400) = \$59.25$ -- which is relatively close to $50.00 and suggests that the point will not greatly affect the regression equation. The following check confirms this.

$\bar{x} = 120.26$

$\bar{y} = 16.23$

$b_1 = [n(\Sigma xy) - (\Sigma x)(\Sigma y)]/[n(\Sigma x^2) - (\Sigma x)^2]$

    $= 81545.01/668584.39$

    $= .1220$

$b_0 = \bar{y} - b_1\bar{x}$

    $= 16.23 - .1220(120.26)$

    $= 1.55$

$\hat{y} = b_0 + b_1x = 1.55 + .1220x$

$n = 7$

$\Sigma x = 841.84$

$\Sigma y = 113.58$

$\Sigma x^2 = 196754.1416$

$\Sigma y^2 = 3309.5364$

$\Sigma xy = 25308.7436$

21. original data

  $n=5$

  $\Sigma x = 4,234,178$

  $\Sigma y = 576$

  $\Sigma x^2 = 3,595,324,583,102$

  $\Sigma y^2 = 67552$

  $\Sigma xy = 491,173,342$

original data divided by 1000

  $n = 5$

  $\Sigma x = 4,234.178$

  $\Sigma y = 576$

  $\Sigma x^2 = 3,595,324.583102$

  $\Sigma y^2 = 67552$

  $\Sigma xy = 491,173.342$

original data

$\overline{x} = 846835.6$
$\overline{y} = 115.2$
$n\Sigma xy - (\Sigma x)(\Sigma y) = 16,980,182$
$n\Sigma x^2 - (\Sigma x)^2 = 48,459,579,826$
$b_1 = 16,980,182/48,459,579,826$
    $= .0003504$
$b_0 = \overline{y} - b_1\overline{x}$
    $= 115.2 - .0003504(846835.6)$
    $= -181.53$
$\hat{y} = b_0 + b_1x$
    $= -181.53 + .0003504x$

original data divided by 1000

$\overline{x} = 846.8356$
$\overline{y} = 115.2$
$n\Sigma xy - (\Sigma x)(\Sigma y) = 16,980.182$
$n\Sigma x^2 - (\Sigma x)^2 = 48,459.579826$
$b_1 = 16,980.182/48,459.579826$
    $= .3504$
$b_0 = \overline{y} - b_1\overline{x}$
    $= 115.2 - .3504(846.8356)$
    $= -181.53$
$\hat{y} = b_0 + b_1x$
    $= -181.53 + .3504x$

Dividing each x by 1000 multiplies b_1, the coefficient of x in the regression equation, by 1000; multiplying the x coefficient by 1000 and dividing x by 1000 will "cancel out" and all predictions remain the same.

Dividing each y by 1000 divides both b_1 and b_0 by 1000; consistent with the new "units" for y, all predictions will also turn out divided by 1000.

23. •original data

x	y
2.0	12.0
2.5	18.7
4.2	53.0
10.0	225.0

$n = 4$
$\Sigma x = 18.7$
$\Sigma y = 308.7$
$\Sigma x^2 = 127.89$
$\Sigma y^2 = 53927.69$
$\Sigma xy = 2543.35$

$b_1 = [n(\Sigma xy) - (\Sigma x)(\Sigma y)]/[n(\Sigma x^2) - (\Sigma x)^2]$
    $= 4400.71/161.87 = 27.2$
$b_0 = \overline{y} - b_1\overline{x}$
    $= (308.7/4) - (27.2)(18.7/4) = -49.9$
$\hat{y} = b_0 + b_1x = -49.9 + 27.2x$
$r = [n(\Sigma xy) - (\Sigma x)(\Sigma y)]/[\sqrt{n(\Sigma x^2) - (\Sigma x)^2} \cdot \sqrt{n(\Sigma y^2) - (\Sigma y)^2}]$
    $= 4400.71/[\sqrt{161.87} \cdot \sqrt{120415.07}]$
    $= .9968$

Based on the value of the associated correlations (.9968 > .9631), the above equation using the original data seems to fit the data better than the following equation using ln(x) instead of x.

•using ln(x) for x

x	y
.693	12.0
.916	18.7
1.435	53.0
2.303	225.0

$n = 4$
$\Sigma x = 5.347$
$\Sigma y = 308.7$
$\Sigma x^2 = 8.681$
$\Sigma y^2 = 53927.69$
$\Sigma xy = 619.594$

$b_1 = [n(\Sigma xy) - (\Sigma x)(\Sigma y)]/[n(\Sigma x^2) - (\Sigma x)^2]$
    $= 827.7220/6.134071 = 134.9$
$b_0 = \overline{y} - b_1\overline{x}$
    $= (308.7/4) - (134.9)(5.347/4) = 103.2$
$\hat{y} = b_0 + b_1 \cdot ln(x) = -103.2 + 134.9 \cdot ln(x)$ [since the "x" is really ln(x)]
$r = [n(\Sigma xy) - (\Sigma x)(\Sigma y)]/[\sqrt{n(\Sigma x^2) - (\Sigma x)^2} \cdot \sqrt{n(\Sigma y^2) - (\Sigma y)^2}]$
    $= 827.7220/[\sqrt{6.134071} \cdot \sqrt{120415.07}]$
    $= .9631$

NOTE: Both x and y (perhaps, especially y) seem to grow exponentially. A wiser choice for a transformation might be to use both ln(x) for x and ln(y) for y (or, perhaps, only ln(y) for y).

9-4 Variation and Prediction Intervals

1. The coefficient of determination is $r^2 = (.3)^2 = .09$.
   The portion of the total variation explained by the regression line is $r^2 = .09 = 9\%$.

3. The coefficient of determination is $r^2 = (-.327)^2 = .107$.
   The portion of the total variation explained by the regression line is $r^2 = .107 = 10.7\%$.

5. Since $r^2 = .928$, $r = +.963$ (positive because $b_1 = 12.5$ is positive).
   From Table A-6, the critical values necessary for significance are $\pm.279$.
   Since $.963 > .279$, we conclude there is a significant positive linear correlation between chest sizes of bears and their weights.

7. 262.74 from the "fit" value on the Minitab display
   NOTE: $\hat{y} = -264.48 + 12.5444x$
   $\hat{y}_{50} = -264.48 + 12.5444(50) = 362.74$

9. The predicted values were calculated using the regression line $\hat{y} = 2 + 3x$.

x	y	$\hat{y}$	$\bar{y}$	$\hat{y}-\bar{y}$	$(\hat{y}-\bar{y})^2$	$y-\hat{y}$	$(y-\hat{y})^2$	$y-\bar{y}$	$(y-\bar{y})^2$
1	5	5	12.2	-7.2	51.84	0	0	-7.2	51.84
2	8	8	12.2	-4.2	17.64	0	0	-4.2	17.64
3	11	11	12.2	-1.2	1.44	0	0	-1.2	1.44
5	17	17	12.2	4.8	23.04	0	0	4.8	23.04
6	20	20	12.2	7.8	60.84	0	0	7.8	60.84
17	61	61	61.0	0	154.80	0	0	0	154.80

   a. The explained variation is $\Sigma(\hat{y}-\bar{y})^2 = 154.80$
   b. The unexplained variation is $\Sigma(y-\hat{y})^2 = 0$
   c. The total variation is $\Sigma(y-\bar{y})^2 = 154.80$
   d. $r^2 = \Sigma(\hat{y}-\bar{y})^2/\Sigma(y-\bar{y})^2 = 154.80/154.80 = 1.00$
   e. $s_e^2 = \Sigma(y-\hat{y})^2/(n-2) = 0/3 = 0$
   $s_e = 0$

NOTE: A table such as the one in the preceding problem organizes the work and provides all the values needed to discuss variation. In such a table, the following must always be true and can be used as a check before proceeding.
   * $\Sigma y = \Sigma\hat{y} = \Sigma\bar{y}$
   * $\Sigma(\hat{y}-\bar{y}) = \Sigma(y-\hat{y}) = \Sigma(y-\bar{y}) = 0$
   * $\Sigma(y-\bar{y})^2 + \Sigma(y-\hat{y})^2 = \Sigma(y-\bar{y})^2$

10. The predicted values were calculated using the regression line $\hat{y} = -187.462 + 11.2713x$.

x	y	$\hat{y}$	$\bar{y}$	$\hat{y}-\bar{y}$	$(\hat{y}-\bar{y})^2$	$y-\hat{y}$	$(y-\hat{y})^2$	$y-\bar{y}$	$(y-\bar{y})^2$
26	90	105.59	273.25	-167.7	28109.3	-15.6	243.1	-183.25	33580.56
45	344	319.75	273.25	46.5	2161.9	24.3	588.2	70.75	5005.56
54	416	421.19	273.25	147.9	21885.7	-5.2	26.9	142.75	20377.56
49	348	364.83	273.25	91.6	8387.2	-16.8	283.3	74.75	5587.56
41	262	274.66	273.25	1.4	2.0	-12.7	160.3	-11.25	126.56
49	360	364.83	273.25	91.6	8387.2	-4.8	23.3	86.75	7525.56
44	332	308.48	273.25	35.2	1240.8	23.5	553.4	58.75	3451.56
19	34	26.69	273.25	-246.6	60790.5	7.3	53.4	-239.25	57240.56
327	2186	2186	2186	0	130964.6	0	1931.9	0	132895.5

   a. The explained variation is $\Sigma(\hat{y}-\bar{y})^2 = 130964.6$

   [There is round-off error calculating $\hat{y}$, the true value is 130963.5.]
   b. The unexplained variation is $\Sigma(y-\hat{y})^2 = 1931.9$

   [There is round-off error calculating $\hat{y}$, the true value is 1932.0.]

c. The total variation is $\Sigma(y-\bar{y})^2 = 132895.5$

d. $r^2 = \Sigma(\hat{y}-\bar{y})^2/\Sigma(y-\bar{y})^2 = 130964.6/132895.5 = .9855$

e. $s_e^2 = \Sigma(y-\hat{y})^2/(n-2) = 1931.9/6 = 322.0$

$s_e = 17.94$

11. The predicted values were calculated using the regression line $\hat{y} = .549270 + 1.47985x$.

x	y	$\hat{y}$	$\bar{y}$	$\hat{y}-\bar{y}$	$(\hat{y}-\bar{y})^2$	$y-\hat{y}$	$(y-\hat{y})^2$	$y-\bar{y}$	$(y-\bar{y})^2$
.27	2	.949	3.25	-2.301	5.295	1.051	1.105	-1.25	1.5625
1.41	3	2.636	3.25	-.614	.377	.364	.133	-.25	.0625
2.19	3	3.790	3.25	.540	.292	-.790	.624	-.25	.0625
2.83	6	4.737	3.25	1.487	2.212	1.263	1.595	2.75	7.5625
2.19	4	3.790	3.25	.540	.292	.210	.044	.75	.5625
1.81	2	3.228	3.25	-.022	.000	-1.228	1.507	-1.25	1.5625
.85	1	1.807	3.25	-1.443	2.082	-.807	.651	-2.25	5.0625
3.05	5	5.063	3.25	1.813	3.286	-.063	.004	1.75	3.0625
14.60	26	26.000	26.00	0	13.837	0	5.663	0	19.5000

a. The explained variation is $\Sigma(\hat{y}-\bar{y})^2 = 13.837$

b. The unexplained variation is $\Sigma(y-\hat{y})^2 = 5.663$

c. The total variation is $\Sigma(y-\bar{y})^2 = 19.500$

d. $r^2 = \Sigma(\hat{y}-\bar{y})^2/\Sigma(y-\bar{y})^2$

$= 13.837/19.500$

$= .7096$

e. $s_e^2 = \Sigma(y-\hat{y})^2/(n-2) = 5.663/6 = .9438$

$s_e = .9715$

13. a. $\hat{y} = 2 + 3x$

$\hat{y}_4 = 2 + 3(4) = 14$

b. $\hat{y} \pm t_{n-2,\alpha/2}s_e\sqrt{1 + 1/n + n(x_o-\bar{x})^2/[n\Sigma x^2-(\Sigma x)^2]}$

$\hat{y} \pm 0$, since $s_e = 0$

The prediction "interval" in this case shrinks to a single point. Since $r^2 = 1.00$ (i.e., 100% of the variability in the y's can be explained by the regression), a perfect prediction can be made. For a practical example of such a situation, consider the regression line $\hat{y} = 1.399x$ for predicting the amount of money y due for purchasing x gallons of gasoline at \$1.399 per gallon -- the "prediction" will be exactly correct every time because of the perfect correlation between the number of gallons purchased and the amount of money due.

15. a. $\hat{y} = .549270 + 1.47985x$

$\hat{y}_{2.50} = .549270 + 1.47985(2.50) = 4.25$

b. preliminary calculations

$n = 8$

$\Sigma x = 14.60$ $\qquad\qquad \bar{x} = 14.60/8 = 1.825$

$\Sigma x^2 = 32.9632$ $\qquad\qquad n\Sigma x^2-(\Sigma x)^2 = 8(32.9632)-(14.60)^2 = 50.5456$

$\hat{y} \pm t_{n-2,\alpha/2}s_e\sqrt{1 + 1/n + n(x_o-\bar{x})^2/[n\Sigma x^2-(\Sigma x)^2]}$

$\hat{y}_{2.50} \pm t_{6,.025}(.9715)\sqrt{1 + 1/8 + 8(2.50-1.825)^2/[50.5456]}$

$4.25 \pm (2.447)(.9715)\sqrt{1.19711}$

4.25 ± 2.60

$1.65 < y_{2.50} < 6.85$ [Fractional values may represent part-time occupancy.]

Exercises 17-20 refer to the chapter problem of Table 9-1. They use the following, which are calculated and/or discussed at various places in the text,

$n = 6$

$\Sigma x = 441.84$ $\hat{y} = -.347279 + .148614x$

$\Sigma x^2 = 36754.1416$ $s_e = 3.26584$

and the values obtained below.

$\bar{x} = (\Sigma x)/n = 441.84/6 = 73.64$

$n\Sigma x^2 - (\Sigma x)^2 = 6(36754.1416) - (441.84)^2 = 25302.2604$

17. $\hat{y}_{50} = -.347279 + .148614(50) = 7.083$

$\hat{y} \pm t_{n-2,\alpha/2}s_e\sqrt{1 + 1/n + n(x_o - \bar{x})^2/[n\Sigma x^2 - (\Sigma x)^2]}$

$\hat{y}_{50} \pm t_{4,.025}(3.26584)\sqrt{1 + 1/6 + 6(50-73.64)^2/[25302.2604]}$

$7.083 \pm (2.776)(3.26584)\sqrt{1.29919}$

7.083 ± 10.334

$-3.25 < y_{50} < 17.42$

$0 \le y_{50} < 17.42$

19. $\hat{y}_{80} = -.347279 + .148614(80) = 11.542$

$\hat{y} \pm t_{n-2,\alpha/2}s_e\sqrt{1 + 1/n + n(x_o - \bar{x})^2/[n\Sigma x^2 - (\Sigma x)^2]}$

$\hat{y}_{80} \pm t_{4,.01}(3.26584)\sqrt{1 + 1/6 + 6(80-73.64)^2/[25302.2604]}$

$11.542 \pm (3.747)(3.26584)\sqrt{1.17629}$

11.542 ± 13.272

$-1.73 < y_{80} < 24.81$

$0 \le y_{80} 24.81$

21. This exercise uses the following values from the chapter problem of Table 9-1, which are calculated and/or discussed at various places in the text,

$n = 6$ $\Sigma x = 441.84$ $b_o = -.347279$ $s_e = 3.26584$

$\Sigma x^2 = 36754.1416$ $b_1 = .148614$

and the values obtained below.

$\bar{x} = (\Sigma x)/n = 441.84/6 = 73.64$

$\Sigma x^2 - (\Sigma x)^2/n = 36754.1416 - (441.84)^2/6 = 4217.0434$

a. $b_o \pm t_{n-2,\alpha/2}s_e\sqrt{1/n + \bar{x}^2/[\Sigma x^2 - (\Sigma x)^2/n]}$

$-.347279 \pm t_{4,.025}(3.26584)\sqrt{1/6 + (73.64)^2/[4217.0434]}$

$-.347279 \pm (2.776)(3.26584)\sqrt{1.45260}$

$-.347279 \pm 10.92667$

$-11.27 < \beta_o < 10.60$

b. $b_1 \pm t_{n-2,\alpha/2}s_e/\sqrt{\Sigma x^2 - (\Sigma x)^2/n}$

$.148614 \pm t_{4,.025}(3.26584)/\sqrt{4217.0434}$

$.148614 \pm (2.776)(3.26584)/\sqrt{4217.0434}$

$.148614 \pm .139608$

$.009 < \beta_1 < .288$

From these results, we conclude that a sample of size n=6 is too small to make a very accurate prediction about the amount of a tip. The "lower limit" for a tip (i.e., the amount of a tip for a bill of $0.00, as indicated by β_o) could be as high as $10.60. At the very least, since the confidence interval for β_1 is completely (albeit barely) above 0.00, we can state that there is a positive correlation between the tip and the amount of the bill.

9-5 Rank Correlation

NOTE: This manual calculates $d = R_x - R_y$, thus preserving the sign of d. This convention means Σd must equal 0 and provides a check for the assigning and differencing of the ranks. In addition, it must always be true that $\Sigma R_x = \Sigma R_y = n(n+1)/2$.

1. In each case the $n = 5$ pairs are pairs of ranks, called R_x and R_y below to emphasize that fact.

a.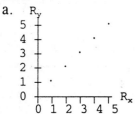

R_x	R_y	d	d^2
1	1	0	0
3	3	0	0
5	5	0	0
4	4	0	0
2	2	0	0
15	15	0	0

```
r_s = 1 - [6(Σd²)]/[n(n²-1)]
    = 1 - [6(0)]/[5(24)]
    = 1 - 0
    = 1
```

Yes; there appears to be a perfect positive correlation between R_x and R_y.

b.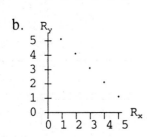

R_x	R_y	d	d^2
1	5	-4	16
2	4	-2	4
3	3	0	0
4	2	2	4
5	1	4	16
15	15	0	40

```
r_s = 1 - [6(Σd²)]/[n(n²-1)]
    = 1 - [6(40)]/[5(24)]
    = 1 - 2
    = -1
```

Yes; there appears to be a perfect negative correlation between R_x and R_y.

c.

R_x	R_y	d	d^2
1	2	-1	1
2	5	-3	9
3	3	0	0
4	1	3	9
5	4	1	1
15	15	0	20

```
r_s = 1 - [6(Σd²)]/[n(n²-1)]
    = 1 - [6(20)]/[5(24)]
    = 1 - 1
    = 0
```

No; there does not appear to be any correlation between R_x and R_y.

3. The following table summarizes the calculations.

R_x	R_y	d	d^2
2	2	0	0
6	7	-1	1
3	7	-3	9
5	4	1	1
7	5	2	4
10	8	2	4
9	9	0	0
8	10	-2	4
4	3	1	1
1	1	0	0
55	55	0	24

```
r_s = 1 - [6(Σd²)]/[n(n²-1)]
    = 1 - [6(24)]/[10(99)]
    = 1 - .145
    = .855
```

H_o: $\rho_s = 0$
H_1: $\rho_s \neq 0$
$\alpha = .05$
C.R. $r_s < -.648$
$r_s > .648$
calculations:
$r_s = .855$
conclusion:

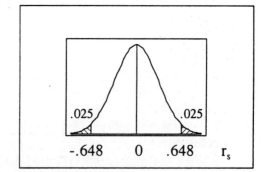

Reject H_o; there is sufficient evidence to reject the claim that $\rho_s = 0$ and to conclude that $\rho_s \neq 0$ (in fact, that $\rho > 0$).

Yes; it does appear that salary increases as stress increases.

IMPORTANT NOTE: The rank correlation is correctly calculated using the ranks in the Pearson product moment correlation formula of chapter 9 to produce

$$r_s = [\Sigma R_x R_y - (\Sigma R_x)(\Sigma R_y)]/[\sqrt{\Sigma R_x^2 - (\Sigma R_x)^2} \cdot \sqrt{\Sigma R_y^2 - (\Sigma R_y)^2}]$$

Since $\Sigma R_x = \Sigma R_y = 1+2+...+n = n(n+1)/2$ [always]

and $\Sigma R_x^2 = \Sigma R_y^2 = 1^2+2^2+...+n^2 = n(n+1)(2n+1)/6$ [when there are ties in the ranks], it can be shown by algebra that the above formula can be shortened to

$$r_s = 1 - [6(\Sigma d^2)]/[n(n^2-1)] \quad \underline{\text{when there are no ties in the ranks}}.$$

As the presence of ties typically does not make a difference in the first 3 decimals of r_s, this manual uses the shortened formula exclusively and notes when use of the longer formula gives a slightly different result.

5. The following table summarizes the calculations.

R_x	R_y	d	d^2
2	5	-3	9
7	2	5	25
6	3	3	9
4	8	-4	16
5	10	-5	25
8	9	-1	1
9	1	8	64
10	7	3	9
3	6	-3	9
1	4	-3	9
55	55	0	176

```
r_s = 1 - [6(Σd²)]/[n(n²-1)]
    = 1 - [6(176)]/[10(99)]
    = 1 - 1.067
    = -.067
```

$H_0: \rho_s = 0$
$H_1: \rho_s \neq 0$
$\alpha = .05$
C.R. $r_s < -.648$
        $r_s > .648$
calculations:
        $r_s = -.067$
conclusion:

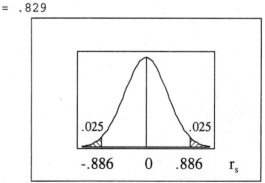

Do not reject H_0; there is not sufficient evidence to conclude that $\rho_s \neq 0$.
No; there does not appear to be a relationship between the stress level of jobs and their physical demands.

7. The following table summarizes the calculations.

x	R_x	y	R_y	d	d^2
33.46	1	5.50	2	-1	1
50.68	2	5.00	2	1	1
87.92	4	8.08	3	1	1
98.84	5	17.00	6	-1	1
63.60	3	12.00	4	-1	1
107.34	6	16.00	5	1	1
	21		21	0	6

```
r_s = 1 - [6(Σd²)]/[n(n²-1)]
    = 1 - [6(6)]/[6(35)]
    = 1 - .171
    = .829
```

$H_0: \rho_s = 0$
$H_1: \rho_s \neq 0$
$\alpha = .05$
C.R. $r_s < -.886$
        $r_s > .886$
calculations:
        $r_s = .715$
conclusion:

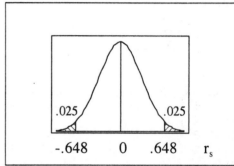

Do not reject H_0; there is not sufficient evidence to conclude that $\rho_s \neq 0$.

9. The following table summarizes the calculations.

x	R_x	y	R_y	d	d^2
17.2	1	.19	3	-2	4
43.5	6	.20	4.5	1.5	2.25
30.7	4	.26	8	-4	16
53.1	8	.16	1	7	49
37.2	5	.24	7	-2	4
21.0	2	.20	4.5	-2.5	6.25
27.6	3	.18	2	1	1
46.3	7	.23	6	1	1
	36.0		36.0	0.0	83.50

$$r_s = 1 - [6(\Sigma d^2)] / [n(n^2-1)]$$
$$= 1 - [6(83.5)] / [8(63)]$$
$$= 1 - .994$$
$$= .006$$

$H_0: \rho_s = 0$
$H_1: \rho_s \neq 0$
$\alpha = .05$
C.R. $r_s < -.738$
        $r_s > .738$
calculations:
    $r_s = .006$

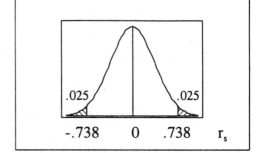

conclusion:
    Do not reject H_0; there is not sufficient evidence to conclude that BAC and age are
    related.

NOTE: Using formula 9-1 (since there are ties) yields $r_s = 0.000$. Compare this to the
parametric hypothesis test using Pearson's correlation in section 9-2 exercise #9.

11. Refer to the data page for exercises #11 and #12. Since there are many ties, use formula
    9-1 applied to the ranks. The critical values are $\pm z/\sqrt{n-1} = \pm 1.96/\sqrt{49} = \pm.280$.
    $n(\Sigma R_i^2) - (\Sigma R_i)^2 = 50(42894.50) - (1275)^2 = 519100$
    $n(\Sigma R_d^2) - (\Sigma R_d)^2 = 50(42917.00) - (1275)^2 = 520225$
    $n(\Sigma R_h^2) - (\Sigma R_h)^2 = 50(42747.00) - (1275)^2 = 511725$
    $n(\Sigma R_i R_d) - (\Sigma R_i)(\Sigma R_d) = 50(40679.25) - (1275)(1275) = 408337.50$
    $n(\Sigma R_i R_h) - (\Sigma R_i)(\Sigma R_h) = 50(32144.75) - (1275)(1275) = -18387.50$

    a. Use the interval (i) & duration (d) values.
        $H_0: \rho_s = 0$
        $H_1: \rho_s \neq 0$
        $\alpha = .05$
        C.R. $r_s < -.280$
              $r_s > .280$
        calculations:
            $r_s = 408337.50/[\sqrt{519100} \sqrt{520225}]$
                $= .7858$

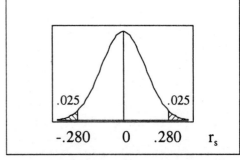

        conclusion:
            Reject H_0; there is sufficient evidence to conclude that the interval between eruptions
            is (positively) correlated with the duration of the last eruption.

    b. Use the interval (i) & height (h) values.
        $H_0: \rho_s = 0$
        $H_1: \rho_s \neq 0$
        $\alpha = .05$
        C.R. $r_s < -.280$
              $r_s > .280$
        calculations:
            $r_s = -18387.50/[\sqrt{519100} \sqrt{511725}]$
                $= -.0356$
        conclusion:
            Do not reject H_0; there is not sufficient evidence to conclude that the interval between
            eruptions is correlated with the height of the last eruption.

Below are the relevant values for exercises #11 and #12.

	exercise #11						exercise #12				
int	R_i	dur	R_d	hgt	R_h	nic	R_n	tar	R_t	car	R_c
86	34.5	240	28	140	30.5	1.2	24.0	16	23.5	15	19.0
86	34.5	237	23.0	154	45	1.2	24.0	16	23.5	15	19.0
62	6.5	122	8	140	30.5	1.0	15.0	16	23.5	17	26.0
104	50	267	40.5	140	30.5	0.8	8.5	9	8.0	6	3
62	6.5	113	4	160	49.5	0.1	1	1	1	1	1
95	45	258	37	140	30.5	0.8	8.5	8	5.5	8	6
79	18.0	232	19	150	40.5	0.8	8.5	10	10	10	8.0
62	6.5	105	3	150	40.5	1.0	15.0	16	23.5	17	26.0
94	44	276	50	160	49.5	1.0	15.0	14	16.5	13	14.0
79	18.0	248	34	155	47.0	1.0	15.0	13	14.0	13	14.0
86	34.5	243	32	125	6.5	1.1	20.0	13	14.0	13	14.0
85	32	241	30.0	136	19.5	1.2	24.0	15	19.0	15	19.0
86	34.5	214	14	140	30.5	1.2	24.0	16	23.5	15	19.0
58	3	114	5	155	47.0	0.7	5.5	9	8.0	11	10.5
89	40.0	272	47	130	10.0	0.9	11	11	11	15	19.0
79	18.0	227	17	125	6.5	0.2	2	2	2	3	2
83	28.0	237	23.0	125	6.5	1.4	28.5	18	28.5	18	28.5
82	24.5	238	25.5	139	23.0	1.2	24.0	15	19.0	15	19.0
84	30.5	203	13	125	6.5	1.1	20.0	13	14.0	12	12
82	24.5	270	44.0	140	30.5	1.0	15.0	15	19.0	16	23.5
78	14.5	218	15	140	30.5	1.3	27	17	27	16	23.5
91	43	226	16	135	15.5	0.8	8.5	9	8.0	10	8.0
89	40.0	250	36	141	37	1.0	15.0	12	12	10	8.0
79	18.0	245	33	140	30.5	1.0	15.0	14	16.5	17	26.0
57	2	120	6.5	139	23.0	0.5	3	5	3	7	4.5
100	48	267	40.5	110	3	0.6	4	6	4	7	4.5
62	6.5	103	2	140	30.5	0.7	5.5	8	5.5	11	10.5
87	37	270	44.0	135	15.5	1.4	28.5	18	28.5	15	19.0
70	11	241	30.0	140	30.5	1.1	20.0	16	23.5	18	28.5
							435.0		435.0		435.0
88	38	239	27	135	15.5						
82	24.5	233	20	140	30.5						
83	28.0	238	25.5	139	23.0						
56	1	102	1	100	1						
81	21.5	271	46	105	2						
74	13	127	10	130	10.0						
102	49	275	48.5	135	15.5						
61	4	140	12	131	12						
83	28.0	264	38	135	15.5						
73	12	134	11	153	43.5						
97	46	268	42	155	47.0						
67	9	124	9	140	30.5						
90	42	270	44.0	150	40.5						
84	30.5	249	35	153	43.5						
82	24.5	237	23.0	120	4						
81	21.5	235	21	138	21						
78	14.5	228	18	135	15.5						
89	40.0	265	39	145	38						
69	10	120	6.5	130	10.0						
98	47	275	48.5	136	19.5						
79	18.0	241	30.0	150	40.5						
	1275.0		1275.0		1275.0						

for exercise #11

$\Sigma R_i^2 = 42894.50$

$\Sigma R_d^2 = 42917.00$

$\Sigma R_h^2 = 42747.00$

$\Sigma R_i R_d = 40679.25$

$\Sigma R_i R_h = 32144.75$

for exercise #12

$\Sigma R_n^2 = 8509.00$

$\Sigma R_t^2 = 8530.00$

$\Sigma R_c^2 = 8519.00$

$\Sigma R_n R_t = 8355.50$

$\Sigma R_n R_c = 7994.25$

c. Based on the results from parts (a) and (b), an eruption's duration is a better predictor of the time interval until the next eruption than is its height. Duration is significantly correlated with interval, but height is not.

13. Refer to the data page for exercises #13 and #14. Since there are many ties, use formula 9-1 applied to the ranks. The critical values for n=30 are ±.364.

$n(\Sigma R_p^2) - (\Sigma R_p)^2 = 30(9455) - (465)^2 = 67425$

$n(\Sigma R_w^2) - (\Sigma R_w)^2 = 30(9452.00) - (465)^2 = 67335$

$n(\Sigma R_c^2) - (\Sigma R_c)^2 = 30(9406.50) - (465)^2 = 65970$

$n(\Sigma R_p R_w) - (\Sigma R_p)(\Sigma R_w) = 30(9079.0) - (465)(465) = 56145$

$n(\Sigma R_p R_c) - (\Sigma R_p)(\Sigma R_c) = 30(6475.5) - (465)(465) = -21960$

a. Use the price (p) & weight (w) values.

$H_o: \rho_s = 0$

$H_1: \rho_s \neq 0$

$\alpha = .05$

C.R. $r_s < -.364$
     $r_s > .364$

calculations:

$r_s = 56145/[\sqrt{67425}\sqrt{67335}]$
    $= .8333$

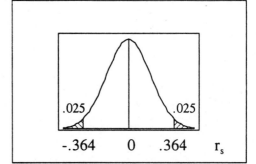

conclusion:

Reject H_o; there is sufficient evidence to conclude that the price of a diamond is (positively) correlated with its weight.

b. Use the price (p) & color (c) values.

$H_o: \rho_s = 0$

$H_1: \rho_s \neq 0$

$\alpha = .05$

C.R. $r_s < -.364$
     $r_s > .364$

calculations:

$r_s = -21960/[\sqrt{67425}\sqrt{65970}]$
    $= -.3293$

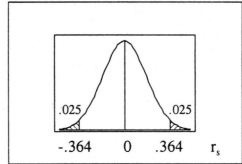

conclusion:

Do not reject H_o; there is not sufficient evidence to conclude that the price of a diamond is correlated with its color rating.

c. A diamond's value is determined more by its weight than by its color. The weight of a diamond is more strongly correlated with its price than is its color. In fact, since the correlation between price and color is not significant, color should ordinarily not be used (at least by itself) when predict or otherwise considering price.

15. a. $t_{6,.025} = 2.365$; $r_s^2 = (2.447)^2/[(2.447)^2 + 6] = .499$, $r_s = ±.707$

b. $t_{13,.025} = 2.160$; $r_s^2 = (2.160)^2/[(2.160)^2 + 13] = .264$, $r_s = ±.514$

c. $t_{28,.025} = 2.048$; $r_s^2 = (2.048)^2/[(2.048)^2 + 28] = .130$, $r_s = ±.361$

d. $t_{28,.005} = 2.763$; $r_s^2 = (2.763)^2/[(2.763)^2 + 28] = .214$, $r_s = ±.463$

e. $t_{6,.005} = 3.707$; $r_s^2 = (3.707)^2/[(3.707)^2 + 6] = .696$, $r_s = ±.834$

Below are the relevant values for exercises #13 and #14.

exercise #13							exercise #14					
pri	R_p	wgt	R_w	col	R_c		gro	R_g	bud	R_b	vie	R_v
6958	11	1.00	1.5	3	8		81.843	16	18.500	11	8.2	28
5885	7	1.00	1.5	5	16.5		194.125	27	140.000	35	6.7	8.5
6333	8	1.01	3.5	4	11.5		147.540	24	50.000	18.5	8.1	27
4299	2	1.01	3.5	5	16.5		75.600	15	72.000	25	8.3	30.5
9589	17	1.02	5	2	4.5		12.006	1	.250	1	7.9	25
6921	10	1.04	6.5	4	11.5		100.853	18	90.000	29.5	8.3	30.5
4426	3	1.04	6.5	5	16.5		67.155	12	104.000	33	6.7	8.5
6885	9	1.07	8.5	4	11.5		140.424	22	75.000	26.5	6.4	4
5826	6	1.07	8.5	5	16.5		68.759	13	55.000	20.5	7.3	13.5
3670	1	1.11	10	9	30		329.691	35	55.000	20.5	7.7	19.5
7176	12	1.12	11	2	4.5		217.631	30	22.000	13	7.1	11
7497	14	1.16	12	5	16.5		198.571	29	3.900	3	8.0	26
5170	4	1.20	13	6	22.5		138.339	21	10.000	7	8.5	34
5547	5	1.23	14	7	26.5		181.280	26	6.000	4	7.3	13.5
18596	24	1.25	15	1	1.5		47.000	5	.325	2	7.7	19.5
7521	15	1.29	16	6	22.5		19.819	2	70.000	23.5	5.2	2
7260	13	1.50	17	6	22.5		72.219	14	17.000	10	6.5	5.5
8139	16	1.51	18	6	22.5		306.124	34	75.000	26.5	6.6	7
12196	21	1.67	19	3	8		197.171	28	39.000	16	7.8	23.0
14998	23	1.72	20	4	11.5		260.000	33	12.000	8	7.8	23.0
9736	18	1.76	21	8	28.5		250.147	32	90.000	29.5	7.4	15
9859	19	1.80	22	5	16.5		20.100	3	45.000	17	6.8	10
12398	22	1.88	23	6	22.5		107.930	20	8.000	6	8.3	30.5
25322	25	2.03	24.5	2	4.5		242.374	31	20.000	12	8.3	30.5
11008	20	2.03	24.5	8	28.5		178.091	25	70.000	23.5	9.1	36
38794	28	2.06	26	2	4.5		96.067	17	25.000	14	8.6	35
66780	30	3.00	27	1	1.5		103.001	19	15.000	9	7.7	19.5
46769	29	4.01	28.5	3	8		48.068	7	110.000	34	4.3	1
28800	26	4.01	28.5	6	22.5		36.900	4	6.400	5	7.7	19.5
28868	27	4.05	30	7	26.5		65.000	11	62.000	22	7.6	16.5
	465		465.0		465.0		63.540	10	90.000	29.5	7.8	23.0
							48.265	8	50.000	18.5	7.6	16.5
							56.876	9	35.000	15	6.5	5.5
							600.743	36	200.000	36	8.4	33
							146.261	23	100.000	32	7.2	12
							47.474	6	90.000	29.5	5.8	3
								666		666.0		666.0

for exercise #13	for exercise #14
$\Sigma R_p^2 = 9455$	$\Sigma R_g^2 = 16206$
$\Sigma R_w^2 = 9452.00$	$\Sigma R_b^2 = 16199.00$
$\Sigma R_c^2 = 9406.50$	$\Sigma R_v^2 = 16192.00$
$\Sigma R_p R_w = 9079.00$	$\Sigma R_g R_b = 12743.50$
$\Sigma R_p R_c = 6475.00$	$\Sigma R_g R_v = 13464.50$

Review Exercises

1. Let x be the price and y be the consumption.

$n = 10$ $n(\Sigma xy) - (\Sigma x)(\Sigma y) = 10(4.74330) - (13.66)(3.468)$
$\Sigma x = 13.66$ $= .06012$
$\Sigma y = 3.468$ $n(\Sigma x^2) - (\Sigma x)^2 = 10(18.6694) - (13.66)^2$
$\Sigma x^2 = 18.6694$ $= .0984$
$\Sigma y^2 = 1.234888$ $n(\Sigma y^2) - (\Sigma y)^2 = 10(1.234888) - (3.468)^2$
$\Sigma xy = 4.74330$ $= .321856$

$r = [n(\Sigma xy) - (\Sigma x)(\Sigma y)]/[\sqrt{n(\Sigma x^2) - (\Sigma x)^2} \cdot \sqrt{n(\Sigma y^2) - (\Sigma y)^2}]$

$\quad = [.06012]/[\sqrt{.0984} \cdot \sqrt{.321856}]$

$\quad = .338$

  a. $H_o: \rho = 0$
    $H_1: \rho \neq 0$
    $\alpha = .05$
    C.R. $r < -.632$ **OR**   C.R. $t < -t_{8,.025} = -2.306$
         $r > .632$ $t > t_{8,.025} = 2.306$

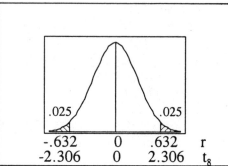

    calculations: calculations:
      $r = .338$ $t_r = (r - \mu_r)/s_r$
                  $= (.338 - 0)/\sqrt{(1-(.338)^2)/8}$
                  $= .338/.333 = 1.015$

    conclusion:
    Do not reject H_o; there is not sufficient evidence to reject the claim that $\rho = 0$.

  b. $r^2 = (.338)^2 = .114 = 11.4\%$

  c. $b_1 = [n(\Sigma xy) - (\Sigma x)(\Sigma y)]/[n(\Sigma x^2) - (\Sigma x)^2]$
      $= .06012/.0984$
      $= .611$
    $b_o = \bar{y} - b_1\bar{x}$
      $= (3.468/10) - (.611)(13.66/10)$
      $= -.488$
    $\hat{y} = b_o + b_1 x$
      $= -.488 + .611x$

  d. $\hat{y} = -.488 + .611x$
    $\hat{y}_{1.38} = \bar{y} = .3468$ pints per capita per week [no significant correlation]

2. Let x be the income and y be the consumption.

$n = 10$ $n(\Sigma xy) - (\Sigma x)(\Sigma y) = 10(1230.996) - (3548)(3.468)$
$\Sigma x = 3548$ $= 5.496$
$\Sigma y = 3.468$ $n(\Sigma x^2) - (\Sigma x)^2 = 10(1259524) - (3548)^2$
$\Sigma x^2 = 1259524$ $= 6936$
$\Sigma y^2 = 1.234888$ $n(\Sigma y^2) - (\Sigma y)^2 = 10(1.234888) - (3.468)^2$
$\Sigma xy = 1230.996$ $= .321856$

$r = [n(\Sigma xy) - (\Sigma x)(\Sigma y)]/[\sqrt{n(\Sigma x^2) - (\Sigma x)^2} \cdot \sqrt{n(\Sigma y^2) - (\Sigma y)^2}]$

$\quad = [5.496]/[\sqrt{6936} \cdot \sqrt{.321856}]$

$\quad = .116$

a. H_o: $\rho = 0$
 H_1: $\rho \neq 0$
 $\alpha = .05$
 C.R. $r < -.632$ OR C.R. $t < -t_{8,.025} = -2.306$
  $r > .632$ $t > t_{8,.025} = 2.306$

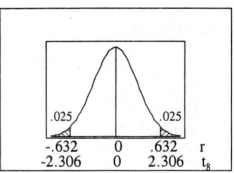

calculations: calculations:
 $r = .116$ $t_r = (r - \mu_r)/s_r$
    $= (.116 - 0)/\sqrt{(1-(.116)^2)/8}$
    $= .116/.351 = .331$

conclusion:
  Do not reject H_o; there is not sufficient evidence to reject the claim that $\rho = 0$.

b. $r^2 = (.116)^2 = .013 = 1.3\%$

c. $b_1 = [n(\Sigma xy) - (\Sigma x)(\Sigma y)]/[n(\Sigma x^2) - (\Sigma x)^2]$
  $= 5.496/6936$
  $= .000792$
 $b_o = \bar{y} - b_1\bar{x}$
  $= (3.468/10) - (.000792)(3548/10)$
  $= .0657$
 $\hat{y} = b_o + b_1 x$
  $= .0657 + .000792x$

d. $\hat{y} = .0657 + .000792x$
 $\hat{y}_{365} = \bar{y} = .3468$ pints per capita per week [no significant correlation]

3. Let x be the temperature and y be the consumption.
  $n = 10$ $n(\Sigma xy) - (\Sigma x)(\Sigma y) = 10(189.038) - (526)(3.468)$
  $\Sigma x = 526$ $= 66.212$
  $\Sigma y = 3.468$ $n(\Sigma x^2) - (\Sigma x)^2 = 10(29926) - (526)^2$
  $\Sigma x^2 = 29926$ $= 22584$
  $\Sigma y^2 = 1.234888$ $n(\Sigma y^2) - (\Sigma y)^2 = 10(1.234888) - (3.468)^2$
  $\Sigma xy = 189.038$ $= .321856$
  $r = [n(\Sigma xy) - (\Sigma x)(\Sigma y)]/[\sqrt{n(\Sigma x^2) - (\Sigma x)^2} \cdot \sqrt{n(\Sigma y^2) - (\Sigma y)^2}]$
  $= [66.212]/[\sqrt{22584} \cdot \sqrt{.321856}] = .777$

a. H_o: $\rho = 0$
 H_1: $\rho \neq 0$
 $\alpha = .05$
 C.R. $r < -.632$ OR C.R. $t < -t_{8,.025} = -2.306$
  $r > .632$ $t > t_{8,.025} = 2.306$

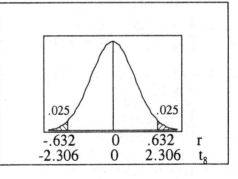

calculations: calculations:
 $r = .777$ $t_r = (r - \mu_r)/s_r$
    $= (.777 - 0)/\sqrt{(1-(.777)^2)/8}$
    $= .777/.223 = 3.486$

conclusion:
  Reject H_o; there is sufficient evidence to reject the claim that $\rho = 0$ and to conclude
  that $\rho \neq 0$ (in fact, $\rho > 0$).

b. $r^2 = (.777)^2 = .604 = 60.4\%$

c. $b_1 = [n(\Sigma xy) - (\Sigma x)(\Sigma y)]/[n(\Sigma x^2) - (\Sigma x)^2]$

    $= 66.212/22584$

    $= .00293$

   $b_0 = \bar{y} - b_1\bar{x}$

    $= (3.468/10) - (.00293)(526/10)$

    $= .193$

   $\hat{y} = b_0 + b_1 x$

    $= .193 + .00293x$

d. $\hat{y} = .193 + .00293x$

   $\hat{y}_{32} = .193 + .00293(32) = .2864$ pints per capita per week

4. The following table summarizes the calculations.

x	R_x	y	R_y	d	d^2
.386	7	41	3	4	16
.374	6	56	5	1	1
.393	8	63	7	1	1
.425	10	68	9	1	1
.406	9	69	10	-1	1
.344	5	65	8	-3	9
.327	4	62	6	-2	4
.288	3	47	4	-1	1
.269	2	32	2	0	0
.256	1	24	1	0	0
	55		55	0	34

$r_s = 1 - [6(\Sigma d^2)]/[n(n^2-1)]$
$= 1 - [6(34)]/[10(99)]$
$= 1 - .206$
$= .794$

$H_0: \rho_s = 0$

$H_1: \rho_s \neq 0$

$\alpha = .05$

C.R. $r_s < -.648$

    $r_s > .648$

calculations:

   $r_s = .774$

conclusion:

   Reject H_0; there is sufficient evidence to reject the claim that $\rho_s = 0$ and to conclude that $\rho_s \neq 0$ (in fact, $\rho_s > 0$).

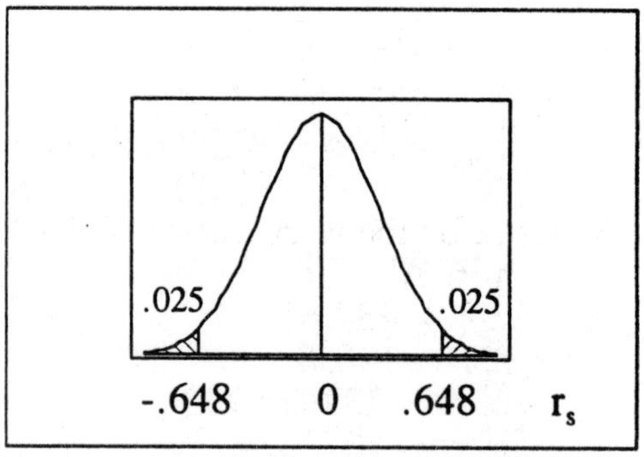

5. Let x be the precipitation and y be the corn production.

$n = 10$

$\Sigma x = 339.1$

$\Sigma y = 14059$

$\Sigma x^2 = 11896.05$

$\Sigma y^2 = 20770345$

$\Sigma xy = 483531.0$

$n(\Sigma xy) - (\Sigma x)(\Sigma y) = 10(483531.0) - (339.1)(14059)$
$= 67903.1$

$n(\Sigma x^2) - (\Sigma x)^2 = 10(11896.05) - (339.1)^2$
$= 3971.69$

$n(\Sigma y^2) - (\Sigma y)^2 = 10(20770345) - (14059)^2$
$= 10047969$

$r = [n(\Sigma xy) - (\Sigma x)(\Sigma y)]/[\sqrt{n(\Sigma x^2) - (\Sigma x)^2} \cdot \sqrt{n(\Sigma y^2) - (\Sigma y)^2}]$

   $= [67903.1]/[\sqrt{3971.69} \cdot \sqrt{10047969}]$

   $= .340$

a. $H_o: \rho = 0$
$H_1: \rho \neq 0$
$\alpha = .05$
C.R. $r < -.632$ **OR** C.R. $t < -t_{8,.025} = -2.306$
$r > .632$ $t > t_{8,.025} = 2.306$

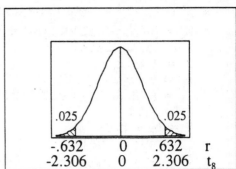

calculations: calculations:
$r = .340$ $t_r = (r - \mu_r)/s_r$
$= (.340 - 0)/\sqrt{(1-(.340)^2)/8}$
$= .340/.333 = 1.022$

conclusion:
Do not reject H_o; there is not sufficient evidence to reject the claim that $\rho = 0$.

b. $b_1 = [n(\Sigma xy) - (\Sigma x)(\Sigma y)]/[n(\Sigma x^2) - (\Sigma x)^2]$
$= 67903.1/3971.69$
$= 17.10$
$b_o = \bar{y} - b_1\bar{x}$
$= 1405.9 - (17.10)(33.91)$
$= 826.1$
$\hat{y} = b_o + b_1x$
$\phantom{\hat{y} }= 826.1 + 17.10x$

c. $\hat{y} = 826.1 + 17.10x$
$\hat{y}_{29.3} = \bar{y} = 1405.9$ million bushels [no significant correlation]

6. Let x be the temperature and y be the corn production.
$n = 10$ $n(\Sigma xy) - (\Sigma x)(\Sigma y) = 10(684765.06) - (487.61)(14059)$
$\Sigma x = 487.61$ $= -7658.39$
$\Sigma y = 14059$ $n(\Sigma x^2) - (\Sigma x)^2 = 10(23799.9197) - (487.61)^2$
$\Sigma x^2 = 23799.9197$ $= 235.6849$
$\Sigma y^2 = 20770345$ $n(\Sigma y^2) - (\Sigma y)^2 = 10(20770345) - (14059)^2$
$\Sigma xy = 68476.06$ $= 10047969$
$r = [n(\Sigma xy) - (\Sigma x)(\Sigma y)]/[\sqrt{n(\Sigma x^2) - (\Sigma x)^2} \cdot \sqrt{n(\Sigma y^2) - (\Sigma y)^2}]$
$= [-7658.39/[\sqrt{235.6849} \cdot \sqrt{10047969}]$
$= -.157$

a. $H_o: \rho = 0$
$H_1: \rho \neq 0$
$\alpha = .05$
C.R. $r < -.632$ **OR** C.R. $t < -t_{8,.025} = -2.306$
$r > .632$ $t > t_{8,.025} = 2.306$

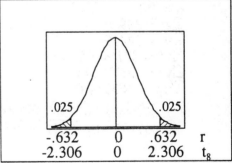

calculations: calculations:
$r = -.157$ $t_r = (r - \mu_r)/s_r$
$= (-.157 - 0)/\sqrt{(1-(-.157)^2)/8}$
$= -.157/.122 = -1.291$

conclusion:
Do not reject H_o; there is not sufficient evidence to reject the claim that $\rho = 0$.

b. $b_1 = [n(\Sigma xy) - (\Sigma x)(\Sigma y)]/[n(\Sigma x^2) - (\Sigma x)^2]$
   $= -7658.39/235.6849$
   $= -32.49$
   $b_0 = \overline{y} - b_1\overline{x}$
   $= 1405.9 - (-32.49)(48.761)$
   $= 2990.3$
   $\hat{y} = b_0 + b_1 x$
   $= 2990.3 - 32.49x$

c. $\hat{y} = 2990.3 - 32.49x$
   $\hat{y}_{48.86} = \overline{y} = 1405.9$ million bushels [no significant correlation]

7. Let x be the acreage and y be the corn production.

n = 10 $n(\Sigma xy) - (\Sigma x)(\Sigma y) = 10(172561550) - (119550)(14059)$
$\Sigma x = 119550$ $= 44862050$
$\Sigma y = 14059$ $n(\Sigma x^2) - (\Sigma x)^2 = 10(1454397500) - (119550)^2$
$\Sigma x^2 = 1454397500$ $= 251772500$
$\Sigma y^2 = 20770345$ $n(\Sigma y^2) - (\Sigma y)^2 = 10(20770345) - (14059)^2$
$\Sigma xy = 172561550$ $= 10047969$

$r = [n(\Sigma xy) - (\Sigma x)(\Sigma y)]/[\sqrt{n(\Sigma x^2) - (\Sigma x)^2} \cdot \sqrt{n(\Sigma y^2) - (\Sigma y)^2}]$
   $= [44862050/[\sqrt{251772500} \cdot \sqrt{10047969}]$
   $= .892$

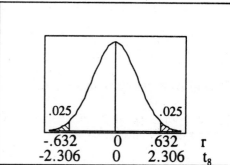

a. $H_0: \rho = 0$
   $H_1: \rho \neq 0$
   $\alpha = .05$
   C.R. $r < -.632$ __OR__   C.R. $t < -t_{8,.025} = -2.306$
         $r > .632$ $t > t_{8,.025} = 2.306$

calculations: calculations:
  $r = .892$ $t_r = (r - \mu_r)/s_r$
                     $= (.892 - 0)/\sqrt{(1-(.892)^2)/8}$
                     $= .892/.160 = 5.579$

conclusion:
   Reject H_0; there is sufficient evidence to reject the claim that $\rho = 0$ and conclude
   that $\rho \neq 0$ (in fact, that $\rho > 0$).

b. $b_1 = [n(\Sigma xy) - (\Sigma x)(\Sigma y)]/[n(\Sigma x^2) - (\Sigma x)^2]$
   $= 44862050/251772500$
   $= .1782$
   $b_0 = \overline{y} - b_1\overline{x}$
   $= 1405.9 - (.1782)(11955.0)$
   $= -724.3$
   $\hat{y} = b_0 + b_1 x$
   $= -724.3 + .1782x$

c. $\hat{y} = -724.3 + .1782x$
   $\hat{y}_{13300} = -724.3 + .1782(13300) = 1645.8$ million bushels

8. The following table summarizes the calculations.

x	R_x	y	R_y	d	d^2
13850	10	1731	10	0	0
13150	8	1578	7	1	1
8550	1	744	1	0	0
12900	7	1445	4	3	9
13550	9	1707	9	0	0
12050	4	1627	8	-4	16
10150	2	1320	3	-1	1
10700	3	899	2	1	1
12250	5	1446	5	0	0
12400	6	1562	6	0	0
	55		55	0	28

$$r_s = 1 - [6(\Sigma d^2)]/[n(n^2-1)]$$
$$= 1 - [6(28)]/[10(99)]$$
$$= 1 - .170$$
$$= -.830$$

H_o: $\rho_s = 0$

H_1: $\rho_s \neq 0$

$\alpha = .05$

C.R. $r_s < -.648$

$\quad r_s > .648$

calculations:

$\quad r_s = .830$

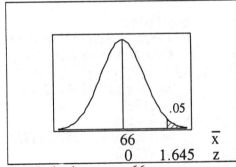

conclusion:

Reject H_o; there is sufficient evidence to reject the claim that $\rho_s = 0$ and to conclude that $\rho_s \neq 0$ (in fact, $\rho_s > 0$).

Cumulative Review Exercises

1. concerns μ: $n > 30$, use z [with s for σ]

summary statistics:

$\quad n = 50$

$\quad \Sigma x = 4033$ $\qquad\qquad \bar{x} = 80.66$

$\quad \Sigma x^2 = 332331$ $\qquad\quad s = 11.98$

a. original claim $\mu > 66$

$\quad H_o$: $\mu \leq 66$

$\quad H_1$: $\mu > 66$

$\quad \alpha = .05$ [assumed]

$\quad$ C.R. z $> z_{.05} = 1.645$

$\quad$ calculations:

$\qquad z_{\bar{x}} = (\bar{x} - \mu)/\sigma_{\bar{x}}$

$\qquad\quad = (80.66 - 66)/(11.98/\sqrt{50}$

$\qquad\quad = 14.66/1.694 = 8.655$

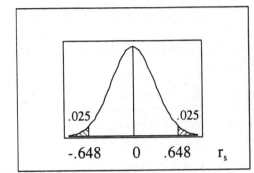

conclusion: reject H_o; there is sufficient evidence to conclude that $\mu > 66$.

b. $\bar{x} \pm z_{.025} \cdot \sigma_{\bar{x}}$

$\quad 80.66 \pm 1.96 \cdot 11.98/\sqrt{50}$

$\quad 80.66 \pm 3.32$

$\quad 77.3 < \mu < 84.0$

2. Let x and y be as given.

$n = 12$

$\Sigma x = 1189$

$\Sigma y = 1234$

$\Sigma x^2 = 118599$

$\Sigma y^2 = 127724$

$\Sigma xy = 122836$

$$n(\Sigma xy) - (\Sigma x)(\Sigma y) = 12(122836) - (1189)(1234)$$
$$= 6806$$
$$n(\Sigma x^2) - (\Sigma x)^2 = 12(118599) - (1189)^2$$
$$= 9467$$
$$n(\Sigma y^2) - (\Sigma y)^2 = 12(127724) - (1234)^2$$
$$= 9932$$

$$r = [n(\Sigma xy) - (\Sigma x)(\Sigma y)]/[\sqrt{n(\Sigma x^2) - (\Sigma x)^2} \cdot \sqrt{n(\Sigma y^2) - (\Sigma y)^2}]$$
$$= [6806]/[\sqrt{9467}\,\sqrt{9932}\,]$$
$$= .702$$

a. $\bar{x} = 1189/12 = 99.1$

   $s_x = \sqrt{9467/(12 \cdot 11)} = 8.47$

b. $\bar{y} = 1234/12 = 102.8$

   $s_y = \sqrt{9932/(12 \cdot 11)} = 8.67$

c. No; there does not appear to be a difference between the means of the two populations. In exploring the relationship between the IQ's of twins, such a two sample approach is not appropriate because it completely ignores the pairings of the scores.

d. Correlation would be appropriate for answering the question "is there a (linear) relationship?"

   $H_o: \rho = 0$

   $H_1: \rho \neq 0$

   $\alpha = .05$ [assumed]

   C.R. $r < -.576$ OR   C.R. $t < -t_{10,.025} = -2.228$
   $\quad\quad r > .576$ $\quad\quad\quad\quad\quad\quad t > t_{10,.025} = 2.228$

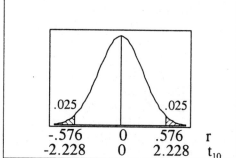

   calculations: calculations:
   $r = .702$ $t_r = (r - \mu_r)/s_r$
   $\quad\quad\quad\quad\quad\quad\quad\quad = (.702 - 0)/\sqrt{(1-(.702)^2)/10}$
   $\quad\quad\quad\quad\quad\quad\quad\quad = .702/.225 = 3.116$

conclusion:

   Reject H_o; there is sufficient evidence to reject the claim that $\rho = 0$ and to conclude that $\rho \neq 0$ (in fact, $\rho > 0$).

Ordinarily, the conclusion would be that about $R^2 = (.702)^2 = .493 = 49.3\%$ of the variation in x can be explained in terms of y (and vice-versa). In this context that means that about 49.3% of the variation among the IQ's in one group can be explained in terms of the IQ's of their twins (i.e, in terms of heredity). In simplest terms, it seems intelligence is about ½ due to heredity and ½ due to environment.

NOTE: The study in exercise #2 contains at least two interesting subtleties.

(1) Correlation addresses only whether there is a relationship between the IQ's, and not whether the IQ's are close to each other. If each older twin had an IQ 20 points higher than the corresponding younger twin, there would be a perfect correlation -- but the twins would not be similar in IQ at all. Beware of the misconception that correlation implies similarity.

(2) Within each pair, the older twin was designated x. That was an arbitrary decision to produce an objective rule, and there is no biological basis for putting all the older twins in one group and the younger twins in another. The x-y designation within pairs could just as properly have been made alphabetically, randomly, or by another rule. But the rule affects the results. If it happens to designate all the twins with the higher IQ as x, the correlation rises to $r = .810$. If it happens to designate the twins with the higher IQ as x in the first six pairs and as y in the last six, the correlation falls to $r = .633$.

One technique which addresses both of the above issues (i.e., it tests for similarity and does not depend upon an x-y designation at all) involves comparing the variation within pairs to the overall variation in IQ's. If there is significantly less variability between twins than there is variability in the general population (or between non-identical twins raised apart), then there is a significant relationship (i.e., in the sense of similarity) between the IQ's of identical twins.

3. a. No; testing for a correlation will not determine whether the position has an effect on the measured values. Adding the 3.00 to each sitting value, for example, would certainly create a position effect. Such a change would affect the regression equation for predicting one value from the other, but it would not affect the correlation. In general, any linear transformation (e.g., measuring in different units) will not effect the correlation -- which is based on the relationship between the standardized scores for each set of scores.

b. The appropriate test would be the μ_d test of the previous chapter -- using the differences between the paired scores.

original claim $\mu_d = 0$ [$n \leq 30$ and σ_d unknown, use t]

$d = x_{sitting} - x_{supine}$: .03 -.64 -.35 .11 .17 .47 .49 .40 .29 .44

$n = 10$

$\Sigma d = 1.41$ $\qquad\qquad\qquad \bar{d} = .141$

$\Sigma d^2 = 1.4727$ $\qquad\qquad s_d = .376$

$H_0: \mu_d = 0$

$H_1: \mu_d \neq 0$

$\alpha = .05$

C.R. $t < -t_{9,.025} = -2.262$

$\qquad t > t_{9,.025} = 2.262$

calculations:

$t_{\bar{d}} = (\bar{d} - \mu_{\bar{d}})/s_{\bar{d}}$

$\qquad = (.141 - 0)/(.376/\sqrt{10})$

$\qquad = .141/.119 = 1.185$

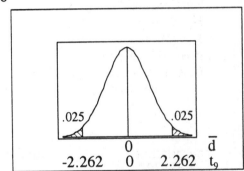

conclusion:

Do not reject H_0; there is not sufficient evidence to reject the claim that $\mu_d = 0$.

Chapter 10

Chi-Square and Analysis of Variance

10-2 Multinomial Experiments: Goodness-of-Fit

1. a. $\chi^2_{37,.10} = 51.805$

   b. $\chi^2_{37,.10} = 51.085 > 38.232 > 29.051 = \chi^2_{37,.90}$

        $.10 < \text{P-value} < .90$

   c. There is not enough evidence to reject the claim that the 38 results are equally likely.

NOTE: In multinomial problems, always verify that $\Sigma E = \Sigma O$ before proceeding. If these sums are not equal, then an error has been made and further calculations have no meaning.

3. $H_o: p_{LF} = p_{RF} = p_{LR} = p_{RR} = .25$

   $H_1:$ at least one of the proportions is different from .25

   $\alpha = .05$

   C.R. $\chi^2 > \chi^2_{3,.05} = 7.815$

   calculations:

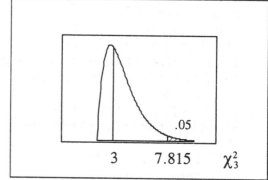

tire	O	E	$(O-E)^2/E$
LF	11	10	.100
RF	15	10	2.500
LR	8	10	.400
RR	6	10	1.600
	40	40	4.600

   $\chi^2 = \Sigma[(O-E)^2/E] = 4.600$

   conclusion:

   Do not reject H_o; there is not sufficient evidence to conclude that at least one of the proportions is different from .25.

   While we cannot be 95% certain that there is a tendency to pick one tire more than any other, it might be worthwhile to take a larger sample to see if the trend toward picking front tires becomes significant.

5. $H_o: p_{Sun} = p_{Mon} = p_{Tue} = \ldots = p_{Sat} = 1/7$

   $H_1:$ at least one of the proportions is different from 1/7

   $\alpha = .05$ [assumed]

   C.R. $\chi^2 > \chi^2_{6,.05} = 12.592$

   calculations:

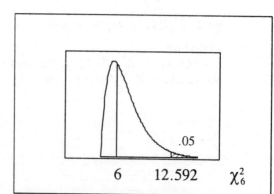

day	O	E	$(O-E)^2/E$
Sun	40	30.857	2.709
Mon	24	30.857	1.524
Tue	25	30.857	1.112
Wed	28	30.857	.265
Thu	29	30.857	.112
Fri	32	30.857	.042
Sat	38	30.857	1.653
	216	216.000	7.417

   $\chi^2 = \Sigma[(O-E)^2/E] = 7.417$

   conclusion:

   Do not reject H_o; there is not sufficient evidence to conclude that at least one of the proportions is different from 1/7.

There is not enough evidence to reject with 95% confidence the daily drinker theory in favor of theory of the casual drinker who binges on Friday and Saturday.

7. H_o: p_{Mon} = .30, p_{Tue} = .15, p_{Wed} = .15, p_{Thu} = .20, p_{Fri} = .20
    H_1: at least one of the proportions is different from what is claimed
    α = .05
    C.R. $\chi^2 > \chi^2_{4,.05}$ = 9.488
    calculations:

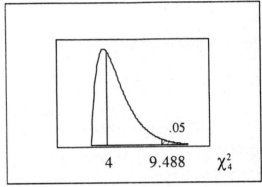

day	O	E	$(O-E)^2/E$
Mon	31	44.10	3.891
Tue	42	22.05	18.050
Wed	18	22.05	.744
Thu	25	29.40	.659
Fri	31	29.40	.087
	147	147.00	23.431

$$\chi^2 = \Sigma[(O-E)^2/E] = 23.431$$

conclusion:
Reject H_o; there is sufficient evidence to conclude that at least one of the proportions is different from what is claimed.

Rejection of this claim may indirectly help to correct the accident problem. It may indicate that the safety expert is wrong and that a better such person should be employed. It may indicate that the safety expert is correct but that Tuesday (which makes an unusually large contribution to $\Sigma[(O-E)^2/E]$) involves circumstances unique to the plant being studied. At the very least, rejection of the hypothesis indicates that medical staffing should not be based on the proportions claimed.

9. H_o: $p_0 = p_1 = p_2 = ... = p_9 = .10$
    H_1: at least one of the proportions is different from .10
    α = .05
    C.R. $\chi^2 > \chi^2_{9,.05}$ = 16.919
    calculations:

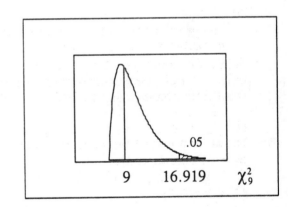

digit	O	E	$(O-E)^2/E$
0	8	10	.400
1	8	10	.400
2	12	10	.400
3	11	10	.100
4	10	10	.000
5	8	10	.400
6	9	10	.100
7	8	10	.400
8	12	10	.400
9	14	10	1.600
	100	100	4.200

$$\chi^2 = \Sigma[(O-E)^2/E] = 4.200$$

conclusion:
Do not reject H_o; there is not sufficient evidence to conclude that at least one of the proportions is different from .10.

11. H_o: $p_1 = .16$, $p_2 = .44$, $p_3 = .27$, $p_4 = .13$
    H_1: at least one of the proportions is different from the license proportions
    $\alpha = .05$
    C.R. $\chi^2 > \chi^2_{3,.05} = 7.815$
    calculations:

group	O	E	$(O-E)^2/E$
1: < 25	36	14.08	34.125
2: 25-44	21	38.72	8.109
3: 45-64	12	23.76	5.821
4: > 64	19	11.44	4.996
	88	88.00	53.051

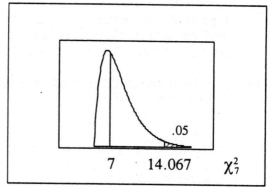

$$\chi^2 = \Sigma[(O-E)^2/E] = 53.051$$

conclusion:
    Reject H_o; there is sufficient evidence to
conclude that at least one of the proportions is different from the license proportions.
Yes; the "under 25" group appears to have a disproportionate number of crashes. It would
be fairer, but much more difficult, to base the E values on the proportion of miles driven
and not on the proportion of licenses possessed.

13. H_o: $p_1 = p_2 = p_3 = ... = p_8 = 1/8$
    H_1: at least one of the proportions is different from 1/8
    $\alpha = .05$ [assumed]
    C.R. $\chi^2 > \chi^2_{7,.05} = 14.067$
    calculations:

start	O	E	$(O-E)^2/E$
1	29	18	6.722
2	19	18	.056
3	18	18	.000
4	25	18	2.722
5	17	18	.056
6	10	18	3.556
7	15	18	.500
8	11	18	2.722
	144	144	16.333

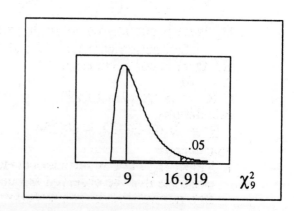

$$\chi^2 = \Sigma[(O-E)^2/E] = 16.333$$

conclusion:
    Reject H_o; there is sufficient evidence to conclude that at least one of the proportions is
different from 1/8 (i.e., that the probabilities of winning in the different starting
positions are not all the same).

15. H_o: $p_0 = p_1 = p_2 = ... = p_9 = .10$
    H_1: at least one of the proportions is different from .10
    $\alpha = .05$
    C.R. $\chi^2 > \chi^2_{9,.05} = 16.919$
    calculations:

digit	O	E	$(O-E)^2/E$
0	33	5.2	148.623
1	10	5.2	4.431
2	3	5.2	.931
3	2	5.2	1.969
4	0	5.2	5.200
5	1	5.2	3.392
6	0	5.2	5.200
7	1	5.2	3.392
8	0	5.2	5.200
9	2	5.2	1.969
	52	52.0	180.308

$$\chi^2 = \Sigma[(O-E)^2/E] = 180.308$$

conclusion:
    Reject H_o; there is sufficient evidence to conclude that at least one of the proportions is different from .10.

No; the reasoning applied in the section example that leads to the conclusion that the values are obtained from estimation rather than actual measurement does not apply here. Since there are so many days with 0.0 precipitation, one would expect a large number of 0's.

17. H_o: $p_{LF} = p_{RF} = p_{LR} = p_{RR} = .25$
    H_1: at least one of the proportions is different from .25
    $\alpha = .05$
    C.R. $\chi^2 > \chi^2_{3,.05} = 7.815$
    calculations:

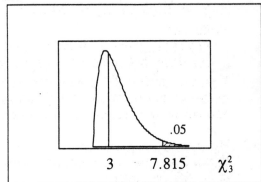

tire	O	E	$(O-E)^2/E$
LF	11	23.5	6.649
RF	15	23.5	3.074
LR	8	23.5	10.223
RR	60	23.5	56.691
	94	94.0	76.638

$$\chi^2 = \Sigma[(O-E)^2/E] = 76.638$$

conclusion:
    Reject H_o; there is sufficient evidence to conclude that at least one of the proportions is different from .25.

In general, an "outlier" is a point that is different from the normal pattern. Since that (i.e., "difference from the normal pattern") is essentially what the goodness-of-fit test is designed to detect, an outlier will tend to increase the probability of rejecting H_o.

19. a. Refer to the illustration at the right.
    $P(x < 79.5) = .5000 - .4147$
        $= .0853$
    $P(79.5 < x < 95.5) = .4147 - .1179$
        $= .2968$
    $P(95.5 < x < 110.5) = .1179 + .2580$
        $= .3759$
    $P(110.5 < x < 120.5) = .4147 - .2580$
        $= .1567$
    $P(x > 120.5) = .5000 - .4147$
        $= .0853$

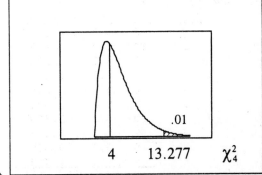

  b.

score	O	E	$(O-E)^2/E$
< 80	20	17.06	.507
80- 95	20	59.36	26.099
96-110	80	75.18	.309
111-120	40	31.34	2.393
>120	40	17.06	30.847
	200	200.00	60.154

  c. H_o: there is goodness of fit to the normal
        distribution with $\mu = 100$ and $\sigma = 15$
    H_1: there is not goodness of fit
    $\alpha = .01$
    C.R. $\chi^2 > \chi^2_{4,.01} = 13.277$
    calculations:
      $\chi^2 = \Sigma[(O-E)^2/E] = 60.154$

    conclusion:
      Reject H_o; there is sufficient evidence to
      conclude that the observed frequencies do not
      fit a normal distribution with $\mu = 100$ and $\sigma = 15$.

10-3 Contingency Tables: Independence and Homogeneity

NOTE: For each row and each column it must be true that $\Sigma O = \Sigma E$. After the marginal row and column totals are calculated, both the row totals and the column totals must sum to produce the same grand total. If either of the preceding is not true, then an error has been made and further calculations have no meaning. In addition, the following are true for all χ^2 contingency table analyses in this manual.
* The E values for each cell are given in parentheses below the O values.
* The addends used to calculate the χ^2 test statistic follow the physical arrangement of the cells in the original contingency table. This practice makes it easier to monitor the large number of intermediate steps involved and helps to prevent errors caused by missing or double-counting cells.
* The accompanying chi-square illustration follows the "usual" shape as pictured with Table A-4, even though that shape is not correct for df=1 or df=2.

1. H_o: drug treatment and oral reaction are independent
   H_1: drug treatment and oral reaction are related
   $\alpha = .05$
   C.R. $\chi^2 > \chi^2_{1,.05} = 3.841$
   calculations:

   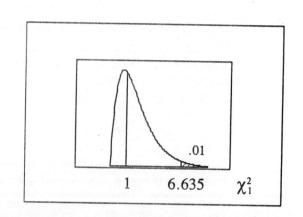

	TREATMENT		
	D	P	
REACTION S	43	35	78
	(38.87)	(39.13)	
N	109	118	227
	(113.13)	(113.87)	
	152	153	305

   $\chi^2 = \Sigma[(O-E)^2/E]$
   $\quad = .4383 + .4355$
   $\quad\quad .1506 + .1496$
   $\quad = 1.174$

   conclusion:
   Do not reject H_o; there is not sufficient evidence to conclude that the drug treatment and the oral reaction are related. NOTE: The P-value from Minitab can also be used to make the decision. Since $.279 > .05$, do not reject H_o.

   A person thinking about using Nicorette might still want to be concerned about mouth soreness. The direction of the data is toward those using the real drug being more likely to experience soreness. While we cannot be 95% sure that there <u>is</u> soreness associated with Nicorette, neither can we be sure that there is <u>not</u> such soreness.

3. H_o: gender and opinion are independent
   H_1: gender and opinion are related
   $\alpha = .01$
   C.R. $\chi^2 > \chi^2_{1,.01} = 6.635$
   calculations:

	OPINION		
	Y	N	
GENDER M	391	425	816
	(457.95)	(358.05)	
F	480	256	736
	(413.05)	(322.95)	
	871	681	1552

   $\chi^2 = \Sigma[(O-E)^2/E]$
   $\quad = 9.7873 + 12.5180$
   $\quad\quad 10.8511 + 13.8787$
   $\quad = 47.035$

conclusion:
   Reject H_o; there is sufficient evidence to reject the claim that gender and opinion are independent and to conclude that these variables are related.

Since more men than women are charged with drunk driving, the men have more to lose.

5. H_o: there is homogeneity of proportions across gender of interviewer
   H_1: there is not homogeneity of proportions
   $\alpha = .01$
   C.R. $\chi^2 > \chi^2_{1,.01} = 6.635$
   calculations:

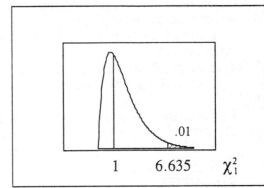

		INTERVIEWER		
		M	F	
FEMALE RESPONSE	A	512 (565.3)	336 (282.7)	848
	D	288 (234.7)	64 (117.3)	352
		800	400	1200

$\chi^2 = \Sigma[(O-E)^2/E]$
$= 5.031 + 10.063$
$\quad 12.121 + 24.242$
$= 51.458$

conclusion:
   Reject H_o; there is sufficient evidence to conclude that there is not homogeneity of proportions across gender of interviewer -- i.e., the proportion of agree/disagree responses vary according to the gender of the interviewer.

7. NOTE: We assume the sample sizes of 200 men and 300 women were predetermined, making this exercise "homogeneity of proportions analysis" rather than "independence of variables."
   H_o: there is homogeneity of proportions across gender of respondent
   H_1: there is not homogeneity of proportions
   $\alpha = .05$
   C.R. $\chi^2 > \chi^2_{2,.05} = 5.991$
   calculations:

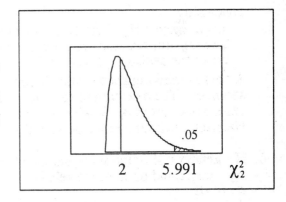

		CONFIDENCE			
		GREAT	SOME	LITTLE	
RESPONDENT	M	115 (116)	56 (60)	29 (24)	200
	F	175 (174)	94 (90)	31 (36)	300
		290	150	60	500

$\chi^2 = \Sigma[(O-E)^2/E]$
$= .009 + .267 + 1.042$
$\quad .006 + .178 + .694$
$= 2.195$

conclusion:
   Do not reject H_o; there is not sufficient evidence to conclude that there is not homogeneity of proportions across gender of respondent -- i.e., the proportion of persons in the given confidence categories does not vary significantly according to the gender of the respondent.

9. H_o: distribution of planes and airline company are independent
    H_1: distribution of planes and airline company are related
    $\alpha = .01$
    C.R. $\chi^2 > \chi^2_{2,.01} = 9.210$
    calculations:

777's	COMPANY			
	U	B	S	
F	36	22	14	72
	(28.15)	(24.36)	(19.49)	
O	16	23	22	61
	(23.85)	(20.64)	(16.51)	
	52	45	36	133

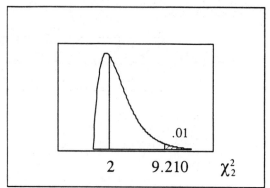

$$\chi^2 = \Sigma[(O-E)^2/E]$$
$$= 2.1888 + .2288 + 1.5458$$
$$2.5835 + .2701 + 1.8246$$
$$= 8.642$$

conclusion:
    Do not reject H_o; there is not sufficient evidence to reject the claim that the distribution of planes is independent of the airline company.

NOTE: This is not evidence to support the given claim, but merely a statement that there is not enough evidence to reject the claim.

11. H_o: type of crime and criminal/victim connection are independent
    H_1: type of crime and criminal/victim connection are related
    $\alpha = .05$
    C.R. $\chi^2 > \chi^2_{2,.05} = 5.991$
    calculations:

CONNECTION	CRIME			
	H	R	A	
S	12	379	727	1118
	(29.93)	(284.64)	(803.43)	
A	39	106	642	787
	(21.07)	(200.36)	(565.57)	
	51	485	1369	1905

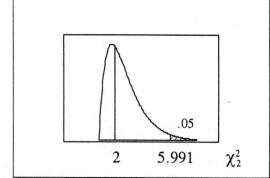

$$\chi^2 = \Sigma[(O-E)^2/E]$$
$$= 10.7418 + 31.2847 + 7.2715$$
$$15.2600 + 44.4425 + 10.3298$$
$$= 119.330$$

conclusion:
    Reject H_o; there is sufficient evidence to conclude that the type of crime and the criminal/victim connection are related.

13. H_o: sentence and plea are independent
H_1: sentence and plea are related
$\alpha = .05$
C.R. $\chi^2 > \chi^2_{1,.05} = 3.841$
calculations:

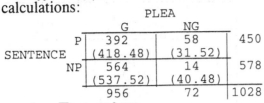

		PLEA		
		G	NG	
	P	392	58	450
SENTENCE		(418.48)	(31.52)	
	NP	564	14	578
		(537.52)	(40.48)	
		956	72	1028

$$\chi^2 = \Sigma[(O-E)^2/E]$$
$$= 1.6759 + 22.2518$$
$$1.3047 + 17.3241$$
$$= 42.557$$

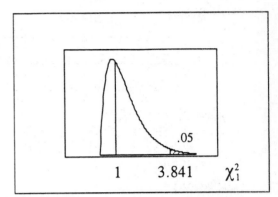

conclusion:

Reject H_o; there is sufficient evidence to conclude that a person's sentence and his original plea are related.

Yes; assuming that those who are really guilty will indeed be convicted with a trial, these results suggest that a guilty plea should be encouraged. But the study reported only those who plead not guilty and were convicted in trials. Suppose there were also guilty 50 persons who plead not guilty and were acquitted. Including them in the no prison category makes the table

		PLEA		
		G	NG	
	P	392	58	450
SENTENCE		(399.07)	(50.93)	
	NP	564	64	628
		(556.93)	(71.07)	
		956	112	1078

$$\chi^2 = \Sigma[(O-E)^2/E]$$
$$= .125 + .982$$
$$.090 + .704$$
$$= 1.901$$

15. H_o: alcohol use and type of crime are independent
H_1: alcohol use and type of crime are related
$\alpha = .05$ [assumed]
C.R. $\chi^2 > \chi^2_{5,.05} = 11.071$

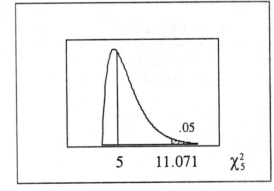

calculations:

		arson	rape	violence	stealing	coining	fraud	
	Y	50	88	155	379	18	63	753
USE		(49.11)	(79.21)	(139.93)	(358.55)	(16.90)	(109.31)	
	N	43	62	110	300	14	144	673
		(43.89)	(70.79)	(125.07)	(320.45)	(15.10)	(97.69)	
		93	150	265	679	32	207	1426

$$\chi^2 = \Sigma[(O-E)^2/E]$$
$$= .016 + 0.976 + 1.622 + 1.167 + .072 + 19.617$$
$$.018 + 1.092 + 1.815 + 1.306 + .080 + 21.949$$
$$= 49.731$$

conclusion:

Reject H_o; there is sufficient evidence to conclude that alcohol use and type of crime are related.

Fraud seems to be different from the other crimes in that it is more like to be committed by someone who abstains from alcohol.

17. H_o: gender and smoking are independent
H_1: gender and smoking are related
$\alpha = .05$ [assumed]
C.R. $\chi^2 > \chi^2_{1,.05} = 3.841$
calculations:

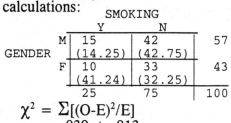

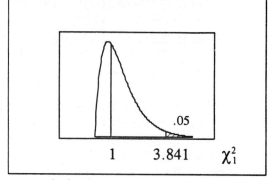

$$\chi^2 = \Sigma[(O\text{-}E)^2/E]$$
$$= .039 + .013$$
$$.052 + .017$$
$$= .122$$

conclusion:
Do not reject H_o; there is not sufficient evidence to conclude that gender and smoking are related.

19. H_o: drug treatment and oral reaction are independent
H_1: drug treatment and oral reaction are related
$\alpha = .05$
C.R. $\chi^2 > \chi^2_{1,.05} = 3.841$
calculations:

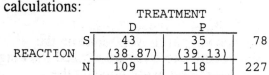

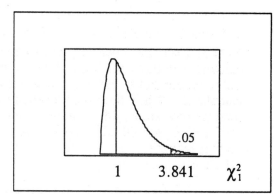

$$\chi^2 = \Sigma[(|O\text{-}E|\text{-}.5)^2/E]$$
$$= .3386 + .3364$$
$$.1163 + .1156$$
$$= .907$$

conclusion:
Do not reject H_o; there is not sufficient evidence to conclude that the drug treatment and the oral reaction are related.

Without the correction for continuity [see the solution for exercise #1 for the details] the calculated test statistic is 1.174. Since $(|O\text{-}E|\text{-}.5)^2 < (O\text{-}E)^2$ whenever $|O\text{-}E| > .25$, Yates' correction generally lowers the calculated test statistic.

10-4 One-Way ANOVA

1. a. $n = $ (total df + 1) = 20 + 1 = 21
b. $F^5_{15} = 3.33$
c. $F^5_{15,.05} = 2.9013$
d. P-value $= .032$
e. The decision can be made using either the F statistic or the P-value.
   • Using F: since 3.33 > 2.9013,
     reject $H_o: \mu_1 = \mu_2 = ... = \mu_6$ and conclude at least one μ_i is different.
   • Using P-value: since .032 < .05,
     reject $H_o: \mu_1 = \mu_2 = ... = \mu_6$ and conclude at least one μ_i is different.
The two methods will always agree with each other.

3. a. $k = 3$, $\sum \overline{x}_i = 294.32$, $\sum \overline{x}_i^2 = 28875.1$

$\overline{\overline{x}} = \sum \overline{x}_i/k$
$= 294.32/3 = 98.107$

$s_{\overline{x}}^2 = \sum(\overline{x}_i - \overline{\overline{x}})^2/(k-1)$
$= [(97.94 - 98.107)^2 + (98.58 - 98.107)^2 + (97.8 - 98.107)^2]/2$
$= .3459/2 = .1729$

$s_{\overline{x}}^2 = [k \cdot \sum \overline{x}_i^2 - (\sum \overline{x}_i)^2]/[k(k-1)]$
$= [3 \cdot 28875.1 - (294.32)^2]/[3(2)]$
$= 1.0376/6 = .1729$

NOTE: Either of the above formulas can be used to calculate $s_{\overline{x}}^2$. While the latter (short-cut) formula was generally preferred in the earlier chapters, we choose to use the former (definition) formula here. Since the ANOVA calculations involve so many steps, using the definition formula helps to keep the calculations more organized and reinforces the concepts involved -- especially when the n_i values are not equal.

$ns_{\overline{x}}^2 = 5 \cdot .1729 = .8647$

b. $s_p^2 = \sum df_i s_i^2/\sum df_i$
$= [4(.568)^2 + 4(.701)^2 + 4(.752)^2]/12$
$= .4598$

NOTE: The manual used the rounded s's provided. Using the original raw data yields a slightly different value.

c. $F = ns_{\overline{x}}^2/s_p^2 = .8647/.4598 = 1.8804$

d. C.V. is $F = 3.8853$ [C.R. is $F > F_{12,.05}^2 = 3.8853$]

e. Using F: since $1.8804 < 3.8853$, fail to reject $H_o: \mu_{18-20} = \mu_{21-29} = \mu_{30+}$.

NOTE: Exercise 3 was worked in complete detail, showing even work done on the calculator without having to be written down. Subsequent exercises are worked showing intermediate steps, but without writing down detail for routine work done on the calculator. While the manual typically shows only three decimal places for the intermediate steps, all decimal places were carried in the calculator. DO NOT ROUND OFF INTERMEDIATE ANSWERS. SAVE calculated values that will be used again. See your instructor or class assistant if you need help using your calculator accurately and efficiently.

NOTE: This section is calculation-oriented. Do not get so involved with the formulas that you miss concepts. This manual arranges the calculations to promote both computational efficiency and understanding of the underlying principles. The following notation is used in this section.

k = the number of groups
n_i = the number of scores in group i (where $i = 1,2,\ldots,k$)
$\overline{x}_i$ = the mean of group i
s_i^2 = the variance of group i
$\overline{\overline{x}} = \sum n_i \overline{x}_i/\sum n_i$ = the (weighted) mean of the group means
$\qquad$ = the overall mean of all the scores in all the groups
$\qquad = \sum \overline{x}_i/k$ = simplified form when each group has equal size n
s_B^2 = the variance between the groups
$\qquad = \sum n_i(\overline{x}_i - \overline{\overline{x}})^2/(k-1)$
$\qquad = n\sum(\overline{x}_i - \overline{\overline{x}})^2/(k-1) = ns_{\overline{x}}^2$ = simplified form when each group has equal size n
s_p^2 = the variance within the groups
$\qquad = \sum df_i s_i^2/\sum df_i$ = the (weighted) mean of the group variances
$\qquad = \sum s_i^2/k$ = simplified form when each group has equal size n
numerator df $= k-1$
denominator df $= \sum df_i$
$\qquad = k(n-1)$ = simplified form when each group has equal size n
$F = s_B^2/s_p^2$ = (variance between groups)/(variance within groups)

5. Since each group has equal size n, the simplified forms can be used. The following preliminary values are identified and/or calculated.

	subcompact	compact	midsize	full size
n	5	5	5	5
Σx	252	265	244	230
Σx^2	12880	14131	11952	10782
$\overline{x}$	50.4	53.0	48.8	46.0
s^2	44.8	21.5	11.2	50.5

$k = 4$

$n = 5$

$s_{\overline{x}}^2 = \Sigma(\overline{x}_i - \overline{\overline{x}})^2/(k-1)$
    $= 8.597$

$\overline{\overline{x}} = \Sigma\overline{x}_i/k$
    $= 49.55$

$s_p^2 = \Sigma s_i^2/k$
    $= 32.000$

$H_o: \mu_1 = \mu_2 = \mu_3 = \mu_4$
$H_1:$ at least one mean is different
$\alpha = .05$
C.R. $F > F_{16,.05}^3 = 3.2389$
calculations:
    $F = ns_{\overline{x}}^2/s_p^2$
    $= 5(8.597)/32.000 = 1.3422$

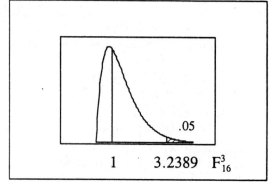

conclusion:
    Do not reject H_o; there is not sufficient evidence to conclude that at least one mean is different.

7. Since each group has equal size n, the simplified forms can be used. The following preliminary values are identified and/or calculated.

	4000 BC	1850 BC	150 AD
n	9	9	9
Σx	1194	1210	1243
Σx^2	158544	162768	171853
$\overline{x}$	132.67	134.44	138.11
s^2	17.500	11.278	22.611

$k = 3$

$n = 9$

$s_{\overline{x}}^2 = \Sigma(\overline{x}_i - \overline{\overline{x}})^2/(k-1)$
    $= 15.416/2 = 7.708$

$\overline{\overline{x}} = \Sigma\overline{x}_i/k$
    $= 135.07$

$s_p^2 = \Sigma s_i^2/k$
    $= 51.389/3 = 17.130$

$H_o: \mu_1 = \mu_2 = \mu_3$
$H_1:$ at least one mean is different
$\alpha = .05$
C.R. $F > F_{24,.05}^2 = 3.4028$
calculations:
    $F = ns_{\overline{x}}^2/s_p^2$
    $= 9(7.708)/17.130 = 4.0498$

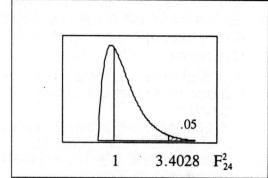

conclusion:
    Reject H_o; there is sufficient evidence to conclude that at least one mean is different.

9. The following preliminary values are identified and/or calculated.

	1	2	3	4	5	total
n	11	11	10	9	7	48
Σx	36.7	39.6	32.5	27.6	25.5	161.9
Σx^2	124.49	144.68	106.73	85.22	94.65	555.77
$\bar{x}$	3.3364	3.6000	3.2500	3.0667	3.6429	3.3729
s^2	.2045	.2120	.1228	.0725	.2929	

$$\bar{\bar{x}} = \Sigma n_i \bar{x}_i / \Sigma n_i$$
$$= [11(3.3364) + 11(3.6000) + 10(3.2500) + 9(3.0667) + 7(3.6429)]/48$$
$$= 161.9/48$$
$$= 3.3729$$

NOTE: This must always agree with the $\bar{x}$ in the "total" column.

$$\Sigma n_i(\bar{x}_i - \bar{\bar{x}})^2 = 11(3.3364-3.3729)^2 + 11(3.6000-3.3729)^2 + 10(3.2500-3.3729)^2$$
$$+ 9(3.0667-3.3729)^2 + 7(3.6429-3.3729)^2$$
$$= 2.087$$

$$\Sigma df_i s_i^2 = 10(.2045) + 10(.2120) + 9(.1228) + 8(.0725) + 6(.2929)$$
$$= 7.608$$

$$s_B^2 = \Sigma n_i(\bar{x}_i - \bar{\bar{x}})^2/(k-1)$$
$$= 2.087/4 = .5217$$

$$s_p^2 = \Sigma df_i s_i^2 / \Sigma df_i$$
$$= 7.606/43 = .1769$$

$H_o: \mu_1 = \mu_2 = \mu_3 = \mu_4 = \mu_5$
$H_1:$ at least one mean is different
$\alpha = .05$
C.R. $F > F_{43,.05}^4 = 2.6060$
calculations:
$$F = s_B^2/s_p^2$$
$$= .5217/.1769 = 2.9491$$

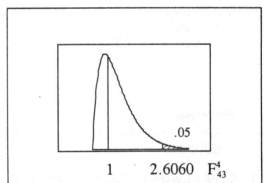

conclusion:

Reject H_o; there is sufficient evidence to conclude that at least one mean is different.

NOTE: An ANOVA table may be completed as follows:

source	SS	df	MS	F
Trt	2.087	4	.522	2.9491
Error	7.608	43	.177	
Total	9.694	47		

$F = MS_{Trt}/MS_{Err}$
$= .522/.177$
$= 2.9491$

(1) Enter $SS_{Trt} = \Sigma n_i(\bar{x}_i - \bar{\bar{x}})^2$ and $SS_{Err} = \Sigma df_i s_i^2$ values from the preliminary calculations.
(2) Enter $df_{Trt} = k-1$ and $df_{Err} = \Sigma df_i = \Sigma(n_i-1) = \Sigma n_i - k$.
(3) Add the SS and df columns to find SS_{Tot} and df_{Tot}. [The df_{Tot} must equal $\Sigma n_i - 1$.]
(4) Calculate $MS_{Trt} = SS_{Trt}/df_{Trt}$ and $MS_{Err} = SS_{Err}/df_{Err}$.
(5) Calculate $F = MS_{Trt}/MS_{Err}$.

As a final check, calculate s^2 (i.e., the variance of all the scores in one large group) two different ways as indicated below. If these answers agree, the problem is probably correct.

* from the "total" column in the table for the preliminary calculations:
$$s^2 = [n\Sigma x^2 - (\Sigma x)^2]/[n(n-1)]$$
$$= [48(555.77)-(161.9)^2]/[48(47)]$$
$$= 465.35/2256$$
$$= .206$$

* from the "total" row of the ANOVA table
$$s^2 = SS_{Tot}/df_{Tot}$$
$$= 9.694/47$$
$$= .206$$

11. The following preliminary values are identified and/or calculated.

	R	O	Y	Br	Bl	G	total
n	21	8	26	33	5	7	100
Σx	19.104	7.401	23.849	30.123	4.507	6.846	91.470
Σx^2	17.934278	6.862429	21.904543	27.546793	4.075729	6.018734	83.802506
$\bar{x}$	.90971	.92513	.91272	.91282	.90140	.92657	.91470
s^2	.000755	.002226	.001144	.001562	.003280	.001499	

$$\bar{\bar{x}} = \Sigma n_i \bar{x}_i / \Sigma n_i$$
$$= [21(.90971) + 8(.92513) + 26(.91272) + 33(.91282) + 5(.90140)$$
$$+ 7(.92657)]/100 = 91.470/100 = .91470$$

NOTE: $\bar{\bar{x}}$ must always agree with the $\bar{x}$ in the "total" column.

$$\Sigma n_i(\bar{x}_i - \bar{\bar{x}})^2 = 21(.90971-.91470)^2 + 8(.92513-.91470)^2 + 26(.91272-.91470)^2$$
$$+ 33(.91282-.91470)^2 + 5(.90140-.91470)^2 + 7(.92657-.91470)^2$$
$$= .00355$$

$$\Sigma df_i s_i^2 = 20(.000755) + 7(.002226) + 25(.001144)$$
$$+ 32(.001562) + 4(.003280) + 6(.001499)$$
$$= .13135$$

$$s_B^2 = \Sigma n_i(\bar{x}_i - \bar{\bar{x}})^2/(k-1)$$
$$= .00355/5 = .00071$$

$$s_p^2 = \Sigma df_i s_i^2 / \Sigma df_i$$
$$= .13135/94 = .001398$$

H_0: $\mu_R = \mu_O = \mu_Y = \mu_{Br} = \mu_{Bl} = \mu_G$
H_1: at least one mean is different
$\alpha = .05$
C.R. $F > F^5_{94,.05} = 2.2899$
calculations:
$$F = s_B^2/s_p^2$$
$$= .00071/.001398 = .5081$$

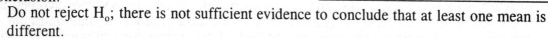

conclusion:

Do not reject H_0; there is not sufficient evidence to conclude that at least one mean is different.

No. If the intent is to make the different colors have the same mean, there is no evidence that this is not being accomplished. Corrective action is not required.

NOTE: An ANOVA table may be completed as follows:

source	SS	df	MS	F
Trt	.00355	5	.00071	.51
Error	.13135	94	.00140	
Total	.13490	99		

$$F = MS_{Trt}/MS_{Err}$$
$$= .00071/.00140$$
$$= .51$$

(1) Enter $SS_{Trt} = \Sigma n_i(\bar{x}_i - \bar{\bar{x}})^2$ and $SS_{Err} = \Sigma df_i s_i^2$ values from the preliminary calculations.
(2) Enter $df_{Trt} = k-1$ and $df_{Err} = \Sigma df_i = \Sigma(n_i-1) = \Sigma n_i - k$.
(3) Add the SS and df columns to find SS_{Tot} and df_{Tot}. [The df_{Tot} must equal $\Sigma n_i - 1$.]
(4) Calculate $MS_{Trt} = SS_{Trt}/df_{Trt}$ and $MS_{Err} = SS_{Err}/df_{Err}$.
(5) Calculate $F = MS_{Trt}/MS_{Err}$.

As a final check, calculate s^2 (i.e., the variance of all the scores in one large group) two different ways as indicated below. If these answers agree, the problem is probably correct.

   * from the "total" column in the table for the preliminary calculations:
   $$s^2 = [n\Sigma x^2 - (\Sigma x)^2]/[n(n-1)]$$
   $$= [100(83.802506) - (91.470)^2]/[100(99)]$$
   $$= 13.4897/9900 = .001363$$

   * from the "total" row of the ANOVA table
   $$s^2 = SS_{Tot}/df_{Tot}$$
   $$= .13490/99$$
   $$= .001363$$

13. a. Adding 2° to each temperature in the 18-20 age group will raise $\bar{x}_1$ by 2° to 99.940, but adding a constant to each value will not affect the spread of the values -- i.e., $s_1 = .568$ does not change. Re-working exercise #3 produces the following.

$$\bar{\bar{x}} = \Sigma \bar{x}_i / k$$
$$= 296.32/3$$
$$= 98.773$$

$$s_{\bar{x}}^2 = \Sigma(\bar{x}_i - \bar{\bar{x}})^2/(k-1)$$
$$= [(99.94-98.773)^2 + (98.58-98.773)^2 + (97.8-98.773)^2]/2$$
$$= 2.346/2$$
$$= 1.1729$$

$$s_p^2 = \Sigma df_i s_i^2 / \Sigma df_i$$
$$= [4(.568)^2 + 4(.701)^2 + 4(.752)^2]/12$$
$$= .4598$$

NOTE: The manual used the rounded s's provided. Using the original raw data yields a slightly different value.

$$F = n s_{\bar{x}}^2 / s_p^2$$
$$= 5(1.1729)/.4598$$
$$= 5.8647/.4598$$
$$= 12.7548$$

C.R. is $F > F_{12,.05}^2 = 3.8853$

since $12.7458 > 3.8853$, reject $H_o: \mu_{18-20} = \mu_{21-29} = \mu_{30+}$ and conclude that at least one of the means is different.

b. For any statistical procedure to be valid, it should not depend upon the units used in the problem. To reach one conclusion if all measurements are made in inches, for example, and another if they are all made in feet would be inconsistent. Accordingly, changing the temperature readings from the Fahrenheit scale to the Celsius scale will not change the calculated F statistic in exercise #3. Mathematically, $C = (5/9) \cdot F - 160/9$. Since the F test is based on variances, and variances are not affected by additive changes, the $-160/9$ has no effect. Multiplying each score by a constant, however, multiplies the variance by the square of that constant. Since both the numerator and the denominator of the calculated F statistic are variances, the multiplicative constant of (5/9) will multiply both the numerator and the denominator of the calculated F ratio by $(5/9)^2$ -- and the calculated F value is not changed.

c. No change. Because adding the same constant c to each score increases all the means by c, each $\bar{x}_i - \bar{\bar{x}}$ difference remains unchanged and s_B^2 is not changed. Because adding the same constant to each score does not affect the spread of the scores, each s_i^2 remains unchanged and s_p^2 is not changed. Since both the numerator and the denominator are unchanged, the calculated F ratio remains the same.

d. No change. Because multiplying each score by the same constant c multiplies all the means by c, each $\bar{x}_i - \bar{\bar{x}}$ difference changes by a factor of c and s_B^2 is changed by a factor of c^2. Because multiplying each score by the same constant c changes standard deviations by a factor of c and variances by a factor of c^2, each s_i^2 and the s_p^2 are changed by a factor of c^2. Since both the numerator and the denominator are changed by a factor of c^2, the calculated F ratio remains the same.

e. No change. Changing the order of the samples will change the order of the addends in the numerator $\Sigma n_i (\bar{x}_i - \bar{\bar{x}})^2/(k-1)$ and the denominator $\Sigma df_i s_i^2 / \Sigma df_i$ of the F ratio. Since changing the order of the addends does not affect the sum, both the numerator and the denominator are unchanged and the calculated F ratio remains the same.

Review Exercises

1. H_o: NRA membership and storage method are independent
   H_1: NRA membership and storage method are related
   $\alpha = .01$
   C.R. $\chi^2 > \chi^2_{1,.01} = 6.635$
   calculations:

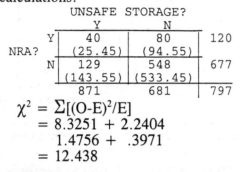

UNSAFE STORAGE?

		Y	N	
NRA?	Y	40 (25.45)	80 (94.55)	120
	N	129 (143.55)	548 (533.45)	677
		871	681	797

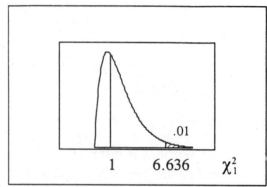

$$\chi^2 = \Sigma[(O-E)^2/E]$$
$$= 8.3251 + 2.2404$$
$$1.4756 + .3971$$
$$= 12.438$$

conclusion:
   Reject H_o; there is sufficient evidence to reject the claim that NRA membership and gun
   storage method are independent and to conclude that these variables are related.

The largest discrepancy (in the Y-Y cell) indicates that NRA members tend to store their
guns loaded and unlocked at a higher rate than non-NRA members. Had the results been
in the opposite direction, it would be logical to credit gun safety programs within the
NRA. While the above results make it difficult to identify reasonable explanatory factors,
some possibilities may be: (1) NRA members tend to live in households where they
suppose the absence of minors doesn't call for safe storage. (2) NRA members tend to
view so-called safe storage as infringing on the right to bear arms.

2. H_o: $p_1 = p_2 = p_3 = ... = p_6 = 1/6$
   H_1: at least one of the proportions is different from 1/6
   $\alpha = .05$
   C.R. $\chi^2 > \chi^2_{5,.05} = 11.071$
   calculations:

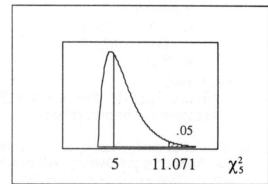

die	O	E	$(O-E)^2/E$
1	2	5	1.800
2	8	5	1.800
3	4	5	.200
4	3	5	.800
5	7	5	.800
6	6	5	.200
	30	30	5.600

$$\chi^2 = \Sigma[(O-E)^2/E] = 5.600$$

conclusion:
   Do not reject H_o; there is not sufficient evidence to conclude that at least one of the
   proportions is different from 1/6.

3. H_o: $p_{Mon} = p_{Tue} = p_{Wed} = \ldots = p_{Sun} = 1/7$
   H_1: at least one of the proportions is different from 1/7
   $\alpha = .05$
   C.R. $\chi^2 > \chi^2_{6,.05} = 12.592$
   calculations:

day	O	E	$(O-E)^2/E$
Mon	74	66.286	.8978
Tue	60	66.286	.5961
Wed	66	66.286	.0012
Thu	71	66.286	.3353
Fri	51	66.286	3.5249
Sat	66	66.286	.0012
Sun	76	66.286	1.4236
	464	464.000	6.7801

   $$\chi^2 = \Sigma[(O-E)^2/E] = 6.780$$

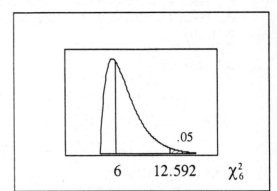

   conclusion:
   Do not reject H_o; there is not sufficient evidence to conclude that at least one of the proportions is different from 1/7.

   The data are very close [calculated $\chi^2 = 6.78$, $E(\chi^2) = 6.00$ if H_o is true] to precisely the amount of random fluctuation we expect if there are no differences among the days and do not support the theory that more gunfire deaths occur on weekends when more people are at home.

4. H_o: there is homogeneity of proportions across airlines
   H_1: there is not homogeneity of proportions
   $\alpha = .05$
   C.R. $\chi^2 > \chi^2_{2,.05} = 5.991$
   calculations:

		AIRLINE			
		USAir	Americ	Delta	
ON	Y	80	77	76	233
TIME?		(77.67)	(77.67)	(77.67)	
	N	20	23	24	67
		(22.23)	(23.33)	(22.33)	
		100	100	100	300

   $$\chi^2 = \Sigma[(O-E)^2/E]$$
   $$= .070 + .006 + .036$$
   $$\quad .244 + .020 + .124$$
   $$= .500$$

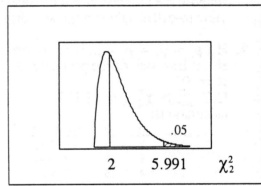

   conclusion:
   Do not reject H_o; there is not sufficient evidence to conclude that there is not homogeneity of proportions.

5. Since each group has equal size n, the simplified forms can be used.
   The following preliminary values are identified and/or calculated.

	oceanside	oceanfront	bayside	bayfront
n	6	6	6	6
Σx	2294	2956	1555	3210
Σx^2	944,422	1,494,108	435,465	1,772,790
$\bar{x}$	382.333	492.667	259.167	535.000
s^2	13469.867	7557.067	6492.167	11088.000

   $k = 4$

   $n = 6$

   $s^2_{\bar{x}} = \Sigma(\bar{x}_i - \bar{\bar{x}})^2/(k-1)$
   $\quad = 45762.243/3$
   $\quad = 15254.081$

   $\bar{\bar{x}} = \Sigma\bar{x}_i/k$
   $\quad = 417.292$

   $s^2_p = \Sigma s^2_i/k$
   $\quad = 38607.100/4$
   $\quad = 9651.775$

H_o: $\mu_{OS} = \mu_{OF} = \mu_{BS} = \mu_{BF}$
H_1: at least one mean is different
$\alpha = .05$
C.R. F > $F^3_{20, .05} = 3.0984$
calculations:
$$F = ns^2_{\bar{x}}/s^2_p$$
$$= 6(15254.081)/9651.775 = 9.4827$$

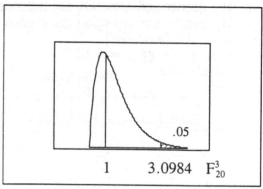

conclusion:
Reject H_o; there is sufficient evidence to conclude that at least one mean is different.

6. H_o: $\mu_A = \mu_B = \mu_C$
H_1: at least one mean is different
$\alpha = .05$
C.R. F > $F^2_{14, .05} = 3.7389$
calculations:
$$F = MS_{Factor}/MS_E$$
$$= .0038286/.0000816 = 46.90$$

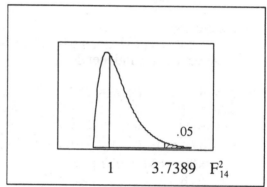

conclusion:
Reject H_o; there is sufficient evidence to reject the claim that the three groups have the same mean level and to conclude that at least one mean is different.

Cumulative Review Exercises

1. scores in order: 66 75 77 80 82 84 89 94;   n = 8 $\Sigma x = 647$ $\Sigma x^2 = 52847$
$\bar{x} = 647/8 = 80.9$
$\tilde{x} = (80 + 82)/2 = 81.0$
$R = 94 - 66 = 28$
$s^2 = [8(52847) - (647)^2]/[8(7)] = 74.41$ $s = 8.6$
five-number summary:
  $x_1 = 66$
  $P_{25} = (75 + 77)/2 = 76.0$ [(.25)8 = 2, a whole number, use $x_{2.5}$]
  $P_{50} = \tilde{x} = 81.0$
  $P_{75} = (84 + 89)/2 = 86.5$ [(.75)8 = 6, a whole number, use $x_{6.5}$]
  $x_8 = 94$

2. Consider the table at the right.
   a. P(C) = 176/647 = .272
   b. P(M) = 303/647 = .468
   c. P(M or C) = P(M) + P(C) - P(M and C)
                  = (303/647) + (176/647) - (82/647)
                  = 397/647
                  = .614
   d. P(F_1 and F_2) = P(F_1)·P(F_2|F_1)
                  = (344/647)·(343/646)
                  = .282

	A	B	C	D	
M	66	80	82	75	303
F	77	89	94	84	344
	143	169	176	159	647

3. H_o: gender and selection are independent
   H_1: gender and selection are related
   $\alpha = .05$ [assumed]
   C.R. $\chi^2 > \chi^2_{3,.05} = 7.815$
   calculations:

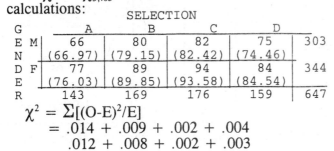

SELECTION

G E M N D F E R	A	B	C	D	
M	66 (66.97)	80 (79.15)	82 (82.42)	75 (74.46)	303
F	77 (76.03)	89 (89.85)	94 (93.58)	84 (84.54)	344
	143	169	176	159	647

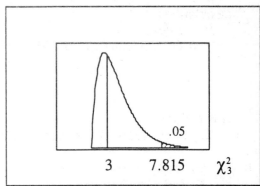

$$\chi^2 = \Sigma[(O-E)^2/E]$$
$$= .014 + .009 + .002 + .004$$
$$.012 + .008 + .002 + .003$$
$$= .055$$

conclusion:
   Do not reject H_o; there is not sufficient evidence to reject the claim that gender and selection are independent.

4. $n = 4$ $\Sigma xy = 26210$ $n(\Sigma xy) - (\Sigma x)(\Sigma y) = 4(26210) - (303)(344) = 608$
   $\Sigma x = 303$ $\Sigma x^2 = 23105$ $n(\Sigma x^2) - (\Sigma x)^2 = 4(23105) - (303)^2 = 611$
   $\Sigma y = 344$ $\Sigma y^2 = 29742$ $n(\Sigma y^2) - (\Sigma y)^2 = 4(29742) - (344)^2 = 632$
   $r = [n(\Sigma xy) - (\Sigma x)(\Sigma y)]/[\sqrt{n(\Sigma x^2) - (\Sigma x)^2} \cdot \sqrt{n(\Sigma y^2) - (\Sigma y)^2}]$
   $= 608/[\sqrt{611} \cdot \sqrt{632}]$
   $= .978$
   original claim: $\rho \neq 0$
   $H_o: \rho = 0$
   $H_1: \rho \neq 0$
   $\alpha = .05$ [assumed]
   C.R. $r < -.950$ **OR**   C.R. $t < -t_{2,.025} = -4.303$
   $r > .950$ $t > t_{2,.025} = 4.303$

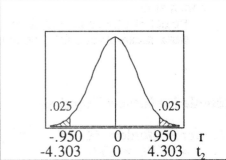

calculations: calculations:
   $r = .978$ $t_r = (r - \mu_r)/s_r$
                        $= (.978 - 0)/\sqrt{(1-(.978)^2)/2}$
                        $= .978/.146 = 6.696$
   conclusion:
   Reject H_o; there is sufficient evidence to conclude that $\rho \neq 0$ (in fact, $\rho > 0$).

5. original claim: $\mu_d > 0$ [$n \leq 30$ and σ_d unknown, use t]
   $d = y - x$: 11 9 12 9;
   $n = 4$
   $\Sigma d = 41$ $\bar{d} = 10.25$
   $\Sigma d^2 = 427$ $s_d = 1.50$
   $H_o: \mu_d \leq 0$
   $H_1: \mu_d > 0$
   $\alpha = .05$ [assumed]
   C.R. $t > t_{3,.05} = 2.353$
   calculations:

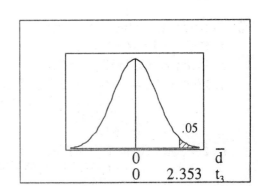

   $t_{\bar{d}} = (\bar{d} - \mu_{\bar{d}})/s_{\bar{d}}$
   $= (10.25 - 0)/(1.50/\sqrt{4})$
   $= 10.25/.750 = 13.667$
   conclusion:
   Reject H_o; there is sufficient evidence to conclude that $\mu_d > 0$.

FINAL NOTE: Congratulations! You have completed statistics – the course that everybody loves to hate. I hope that this manual has helped to make the course a little more understandable for you. I hope you leave the course with an appreciation of broad principles and not memories of merely manipulating formulas. I wish you well in your continued studies, and that you achieve your full potential wherever your journey of life may lead.